INTEGRATED GOVERNANCE AND WATER BASIN MANAGEMENT
Conditions for Regime Change and Sustainability

ENVIRONMENT & POLICY

VOLUME 41

The titles published in this series are listed at the end of this volume.

Integrated Governance and Water Basin Management

Conditions for Regime Change and Sustainability

Edited by

Hans Bressers

University of Twente,
Enschede, The Netherlands

and

Stefan Kuks

University of Twente,
Enschede, The Netherlands

The EUWARENESS research project, "European Water Regimes and the Notion of a Sustainable Status", has been funded by:

EUROPEAN COMMISSION
RESEARCH DIRECTORATE-GENERAL

5th Framework Programme

Under contract-number EVK1-CT-1999-0038

SPRINGER SCIENCE+BUSINESS MEDIA, B.V.

A C.I.P. Catalogue record for this book is available from the Library of Congress.

ISBN 978-90-481-6666-4 ISBN 978-1-4020-2482-5 (eBook)
DOI 10.1007/978-1-4020-2482-5

Printed on acid-free paper

Contents

Preface

This book is a product of the research project EUWARENESS: European Water Regimes and the Notion of a Sustainable Status (EVK1-CT-1999-0038), which represents a two-year cooperation between research institutes from six European countries (2000-2002). The project was financially supported by the European Commission under the 5th Framework Programme, and coordinated by the University of Twente in the Netherlands (see also www.euwareness.nl). The EUWARENESS project aimed to develop a better understanding of the dynamic relationships between various conflicting uses of water resources, the regimes under which these uses of water resources are managed, and conditions generating regime shifts towards sustainability. Important questions in this regard have been: Do integrated water regimes lead to more sustainable water use? What are important indicators for integrated water regimes? Under what conditions can integrated water regimes be achieved? What might be the influence of European and national conditions, to achieve regime transitions at water basin scale? The EUWARENESS project started to study the long-term evolution of national regimes in six European countries (Netherlands, Belgium, France, Spain, Italy, Switzerland). These country studies are published by Kluwer in another book, together with a survey of the evolution of European Water Policy.[1] The long-term regime studies were followed by two case studies per country, in which more specific regime transitions at water basin level are analysed and described. We focused the case studies on the last three decades of the past century (1970-2000), a period during which all countries have attempted to achieve integrated water

[1] See: Kissling-Näf, Ingrid; Kuks, Stefan M.M. (eds.) (2004), The Evolution of National Water Regimes in Europe. Transitions in water rights and water policies towards sustainability. Dordrecht-Boston-London: Kluwer Academic Publishers.

management. In this book we report on these twelve case studies. Rather than just including the case studies as separate chapters, we have chosen to write country-based chapters, in which the two case studies are summarized and discussed within their national context. Furthermore, the book includes a comparative analysis of all case studies, framed in terms of conditions that are important for regime change towards sustainability.

The European comparison in this book would not have been possible without the extensive work carried out by our colleagues in the six European countries studied. They conducted the detailed empirical work on which this book is based and we would like to express our gratitude for this. We also wish to thank the European Commission for funding this project and the preparation of this book, the content of which, however, does not represent its views and in no way anticipates the Commission's future policy in the field of water management.

While being fully aware that much research work remains to be done in this area, we hope that this book will make an interesting contribution to the debate surrounding the governance and sustainable use of water as a resource.

Hans Bressers and Stefan Kuks
November 2003

List of Contributors

David Aubin is researcher in environmental policy analysis at the Department of Political and Social Sciences, Catholic University of Louvain, Belgium. Web-site: www.aurap.ucl.ac.be E-mail: aubin@spri.ucl.ac.be

Hans Bressers is professor of policy studies and environmental policy and scientific director of the Center for Clean Technology and Environmental Policy (CSTM) at the University of Twente, the Netherlands. Web-site: www.utwente.nl/cstm E-mail: j.t.a.bressers@utwente.nl

Meritxell Costejà is a PhD researcher at the Universitat Autònoma of Barcelona. She has participated in several international and national research projects on environmental policy and natural resource management.

Bruno Dente is professor of public policy analysis at the Faculty of Architecture - Urban Planning - Environment of the Milano Politechnic. Web-site: www.polimi.it E-mail: bruno.dente@polimi.it

Jean Marc Dziedzicki holds a PhD in spatial planing and is currently expert for collaborative processes at Réseau Ferré de France. E-mail: jean-marc.dziedzicki@rff.fr

Nuria Font is professor of political science at the Department of Political Science and Public Law at the Universitat Autònoma of Barcelona, Spain. E-mail: nuria.font@uab.es

Doris Fuchs is assistant professor of political science at the University of Munich. She is also affiliated as a senior researcher with the Center for Clean Technology and Environmental Policy at the University of Twente. E-mail: doris.fuchs@lrz.uni-muenchen.de

Alessandra Goria is senior researcher in energy and environmental economics at Fondazione Eni Enrico Mattei (FEEM) in Milano. Web-site: www.feem.it E-mail: alessandra.goria@feem.it

Dave Huitema is a senior researcher at the Institute of Environmental Studies, Vrije Universiteit Amsterdam, the Netherlands. Web-site: www.vu.nl/ivm E-mail: dave.huitema@ivm.vu.nl

Peter Knoepfel is ordinary professor in policy analysis and environmental policies at the Swiss Graduate School for Public Administration (IDHEAP) in Lausanne. Web-site: www.idheap.ch.

Stefan Kuks is senior researcher in comparative water policy studies at the University of Twente (Center for Clean Technology and Environmental Policy) in the Netherlands. He is also executive committee member of the regional water authority "Waterschap Regge en Dinkel". Web-site: www.utwente.nl/cstm E-mail: s.m.m.kuks@utwente.nl

Corinne Larrue is professor in environmental public policies and spatial planning at the University of Tours, France. E-mail: corinne.larrue@univ-tours.fr

Corine Mauch is researcher and consultant in environmental policy issues at Interface Institute for Policy Studies in Lucerne, Switzerland. She is also affiliated as a researcher in a doctoral thesis with the Institut des hautes études en administration public (IDHEAP) in Lausanne. Web-site: www.interface-politikstudien.ch E-mail: mauch@interface-politikstudien.ch

Emmanuel Reynard is assistant professor in physical geography at the Institute of Geography of the University of Lausanne (Switzerland). Website: www.unil.ch/igul/ E-mail: Emmanuel.Reynard@igul.unil.ch

Joan Subirats is professor of political science at the Universitat Autònoma de Barcelona, Spain and Director of the Institute of Government and Public Policy at the same university. E-mail: joan.subirats@uab.es

Frédéric Varone is professor in comparative policy analysis at the Department of Political and Social Sciences, catholic University of Louvain, Belgium. Web-site: www.aurap.ucl.ac.be E-mail: varone@spri.ucl.ac.be

Chapter 1

Governance of Water Resources

Introduction

Hans Bressers and Stefan Kuks
University of Twente (Enschede-Netherlands)

This book is based on the results of the European research project EUWARENESS: European Water Regimes and the Notion of a Sustainable Status. In this project we focused on the sustainable use of water resources, to be achieved by means of integrated water management. Our aim is to develop a better understanding of the dynamic relationships between various conflicting uses of water resources, the regimes under which these uses of water resources are managed, and conditions generating regime shifts towards sustainability. The editors of this book have coordinated the project, in which research groups from six European countries (Netherlands, Belgium, France, Spain, Italy and Switzerland) participated. For these six countries we studied the long-term evolution of national regimes over a period of more than a hundred years. We also studied in greater depth the specific regime transitions of two selected water basins in each country during the last three decades. Important questions have been:

- Do integrated water regimes lead to more sustainable water use?
- What are important indicators for integrated water regimes?
- Under what conditions can integrated water regimes be achieved?
- What could be the influence of Europe and national conditions, to achieve regime transitions at water basin scale?

Hans Bressers and Stefan Kuks (eds.), Integrated governance and water basin management. Conditions for regime change towards sustainability, 1-21. © 2004 Kluwer Academic Publishers.

1.1 Resource access, rivalries and property domains

Being a research project in the field of water policy studies, the study had specific characteristics. We looked at the accessibility of water systems as a natural resource for various users and use functions. In that context we considered rivalries between users and use functions as an indicator of an insufficiently sustainable use of water systems. A water system means a discrete and homogeneous element of surface water or groundwater such as an aquifer, a lake, a reservoir, a stretch of stream, river or canal, an estuary, or a stretch of coastal water. We assume that the sustainable use of water systems requires an optimum distribution of use options among present and future users and use functions. As an example of distribution of use options one could think of the distribution between upstream and downstream users. An activity that pollutes water upstream (using a stream to discharge waste or waste water) could interfere with the downstream use of that stream for drinking water supply. Or an upstream weir could impede the downstream flow and flow dependent use options. Such rivalries not only exist between different (heterogeneous) use types, they may also appear among homogeneous uses (uses of the same type). In arid areas farmers may feel the need to co-ordinate the water use for irrigation. Or in the field of fisheries, quotas may be used as an instrument to prevent the depletion of fish stocks.

A water system is often demarcated as a river or water basin, which means the area of land from which all surface water run-off flows through a sequence of streams, rivers and, possibly, lakes into the sea via a single river mouth, estuary or delta. This implies that a water basin not only includes the water beds, but also the surrounding area of land from which the water bed receives and transports the water run-off. In this view the land use of river flood plains for urban development should be considered as a use that interferes with the use of flood plains for river dynamics and flooding. Another example of a water use rivalry in a water basin could be the rivalry between drainage of land for agricultural development versus the function of a minimum groundwater level for nature conservation in that area.

The main question for the researchers in this project was whether the regime for the management of a water system provided sufficient guarantees for its sustainable use, by diminishing or preventing rivalries between users and use functions. To answer this question we focused on institutional regimes for natural resources, both from a public governance perspective (Bressers and Kuks, 2001) and a perspective of private property and use rights (Ostrom, 1990; Bromley, 1991). The first perspective focuses on the management of natural resources from a public domain (although in interaction with private actors). The second perspective focuses on the accessibility of a natural resource in a broader sense, including the private domain, the domain of collective property and use, as well as the domain of 'no property' (res nullius). By applying both perspectives in a complementary way, we have developed a framework for understanding the access rights that users or use functions may possess or claim, and the proportion between and exclusiveness of the various domains. For instance, the option of having intervention from the public domain could be blocked by the existence of a private domain based on long term concessions for water use (which, especially in Spain and Switzerland, appears to be a problem of redistributing water access rights). On the other hand, attempts initiated by the public domain to redistribute private property and use rights could be effective in providing a better access to or protection of alternative users and use functions. Another question could be how the exclusiveness of the public domain is interpreted by public authorities. Does the public domain offer equal access to society as a whole, or are specific users and/or use functions discriminated against in favour of others? A public domain could appear as a private domain in the hands of society at large, or as a 'no property' domain, owned by nobody and thus equally accessible to everyone.

1.2 Analysis of institutional regimes and regime evolution

In the Euwareness study, a theoretical framework of institutional regime analysis is used that combines property rights theory and institutional rational choice with approaches from political science (policy analysis, in particular policy design theory), thereby innovating the theory of institutional regime.[1] This study examines the potential for an increased integration of institutional elements, such as property rights to natural resources, and the deliberate combination of these institutional elements with other resource protection and use policies. So far, the response to the deterioration in water resource quality arising from the effects of economic growth has taken the form of environment policy intervention. However, the

[1] This framework has its roots in a Swiss project developed by Peter Knoepfel, Ingrid Kissling-Näf and Frédéric Varone, entitled "Comparative analysis of the formation and outcomes of resource regimes in Switzerland", funded by the Swiss National Science Foundation and carried out in the period 1999–2002. For more information see www.idheap.ch, Knoepfel, Kissling-Näf and Varone (2000, 2001) and Kissling-Näf and Varone (2000a, 2000b). Most of section 2 in this chapter is derived from their work. A further elaboration is developed in the EUWARENESS study (see chapter 2 of this book).

capacity for government intervention in this area of the environment is limited due to the existence of implementation deficits, the restriction of traditional environmental policy to selective and often individual and media-oriented emissions management, and the frequent absence of an integrated management of water resources. We define integrated management of water resources as conscious, planned management that takes account of the joint impacts of all forms of use of a given water resource. This means that the interaction between public policies on the one hand and property rights on the other should be taken into account, and that management options for redistribution of property rights should be considered and utilized.

The framework which we developed incorporates the ideas from the theory of property and use rights (especially those of Ostrom and Bromley). However, this theory has its limitations. There are at least three considerations that necessitate the development of a wider regime concept.

While Ostrom fails to take account of the possibilities for public intervention, we have added the political steering dimension to our approach. Ostrom's approach focuses on common-pool resources and -- particularly in the earlier studies of irrigation -- is based on the assumption of a homogenous demand for local commodities and services. In such cases it was possible to prevent the degradation of resources on the basis of voluntary co-operation, i.e. without state intervention. Although, from an economic perspective, this can be viewed as a very efficient strategy, this kind of solution is probably uncommon in highly developed societies characterized by increasingly heterogeneous demands and an expanding scope of effects -- factors which militate against a local and regional solutions like common property. Thus, guidance of heterogeneous, growing and increasingly rivalrous use demands is required.

State regulation of the production and/or consumption of certain goods and services of a natural resource is a common occurrence in everyday political life. However, consideration of the distribution of property and use rights alone is not sufficient for the analysis of the institutional framework. In most cases, there are several public policies which regulate the use of a resource and which, as a result of insufficient co-ordination, can result in the degradation of that resource. In fact, one should consider the influences of all other public policies on a specific commodity or the entire resource.

Unlike Ostrom and Bromley, we do not consider institutions as a framework within which actions take place but as the result and an integral part of the political process. In our opinion, institutions should not merely be understood as frameworks within which actions are carried out. They too are the product and integral components of the political process. Most of the literature concentrates on the analysis of regimes as they exist today. Less emphasis is placed on the perspective dealing with analysis of the process of regime evolution. In order to avoid further degradation of resources, it is important to know when and under what conditions in the political process the regimes can be changed and how this can be brought about and managed.

Institutions are usually understood as sets of rules which structure the relationship between individuals by determining their range of actions in certain situations. Institutions are both the result of past actions and the framework within which new activities take place. Institutions, and hence regimes, can change over time and become increasingly differentiated. In this study the resource regimes have been identified and classified according to a historic approach. The historical emergence of an institutional resource regime can be divided very schematically into different phases (see Figure 1.1).

Figure 1.1: *The growing scarcity or degradation of a resource and development of a regime*

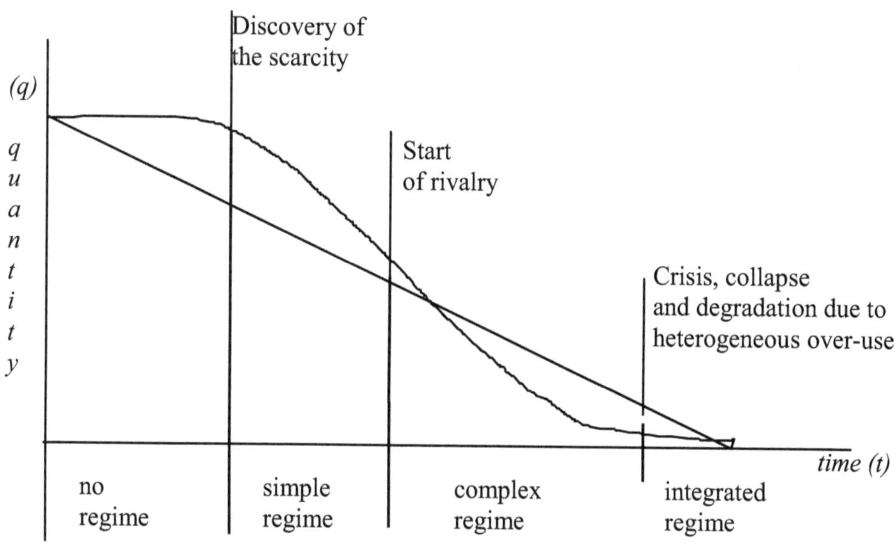

Source: Swiss study by Knoepfel, Kissling-Näf and Varone, see footnote 1.

When the central actors observe scarcity in connection with the use of the resource, this gives rise to a *simple* regime with the new application of property and use rights, or on the policy level with the adoption of general police policies with very simple designs (for instance, a general police clause for protection of use rights or bans and license restrictions).

In the subsequent phase of the *complex* regime, we can already observe the combination of the (clarified, redefined) property and use rights with a more detailed policy, framed in terms of aims of the corresponding protection and use policies. This is followed by a significant increase in the officially sanctioned uses -- which are thus guaranteed in a property-like manner -- for the production of specific goods and services. Examples of such rival uses of water resources are fishing, energy production, agriculture, biodiversity, drinking water production, and recreation. The now mainly heterogeneous demands and the sum of the diverse private-use rights lead to a crisis in and possibly even the collapse of complex regimes.

The key question now is whether it is possible to establish a water resource regime that can take account of these varied, heterogeneous demands. Such regimes could regulate the totality of uses in such a way that it is possible to sustainably maintain the capacity of the water resource in question to the satisfaction of all these uses and use functions *(integrated regime)*. The development of such a regime would depend on institutional changes in the governance structure and the structure of property and use rights.

In the Euwareness study, and thus in this book, we aimed to examine when, whether, under what conditions and in what form integrated water resource regimes are established that can successfully regulate all of the use demands and thus react to the growing scarcity of goods and services provided by a water resource or the depletion of its stocks. A historical screening of water resource regimes in the different countries is used to examine whether the emergence of resource regimes corresponds to our theoretical phase model and whether the suggested transitions from complex to integrated regimes can be identified. This development of national regimes has been examined for a period extending over more than a hundred years (from the early 19th century until the end of the 20th century). The methodology was descriptive, making as much use as possible of previous studies and publications. The screening concentrated on changes in the central elements of the governance structure and the structure of property and use rights. This diachronic analysis made it possible to pronounce on the extent of the regime and on changes in the governance structure and the structure of property and use rights over time. It is particularly important that the transitions are identified, i.e. those historical moments when the resource regime actually changed from one phase to another. In a separate book we report on the historical screening of water resource regimes in the six countries involved (see Kissling-Näf and Kuks, 2004).

Since we are particularly interested in the change from complex to integrated regimes, we subsequently studied two cases of water basins in each of the countries involved, to get a better understanding of the specific conditions under which regime transitions towards integration appear. In this book we report on those twelve case studies of water basins across Europe. The case studies focus on the last three decades of the twentieth century (1970-2000), when all countries have attempted to achieve integrated water management. This forms an important contrast with the other book on national regime studies, since this book focuses on specific examples of regime transitions in recent times.

1.3 Institutional criteria for sustainable water management

1.3.1 Criteria from the EU Water Framework Directive

The dependent variable in our study is the sustainability of water resource regimes. Since our study had to contribute to the implementation of the European Water Framework Directive, we prefer to stay close to the way in which a sustainable status of water resources (water bodies, water basins) is defined in this directive. Key aims of the directive are the following (EU, 2000):

- expanding the scope of water protection to all waters: surface waters and groundwater;
- achieving a 'good status' for all waters by a set deadline;
- water management based on a river basin approach;
- 'combined approach' of emission limit values and quality standards;
- getting the prices right;
- getting the citizen involved more closely;
- streamlining legislation.

Considering these aims, we could distinguish between criteria for ecological sustainability and for sustainable management or institutional sustainability. Ecological sustainability refers to the achievement of a 'good status' for all waters by a set deadline. It addresses the status of water quality and water quantity in terms of its availability and fitness for present and future demands (uses). There are many European directives that set standards for such a good status. The European Water Framework Directive aims to integrate these standards. Ecological sustainability especially aims to protect the way in which water resources are needed for the ecosystem (ecological uses) or for human health (such as the Drinking Water Directive).

In our view, institutional sustainability is a derivative of ecological sustainability. It focuses on the availability and fitness of water resources for all interested users, user groups and use functions that could be identified. Institutional sustainability requires the distribution of a resource among users in a just manner, also taking into account the availability and reproductive capacity of a resource for future generations. It requires from water management that it recognises users and use functions and provides them with rights of access to a water resource (property rights). Additionally, the concept of institutional sustainability provides criteria for resource regulation by means of public policies. Such criteria concern how we would design a water policy or management regime to finally improve the ecological sustainability of water use. This refers to water management at basin level, the participation of users in policy making, expanding the policy scope to all aspects of a water resource, streamlining legislation, getting prices right and redistributing access rights among users.

1.3.2 Ecological sustainability based on good status criteria[2]

The European Water Framework Directive focuses on the good status of water, while addressing both quality and quantity aspects of water resources. There are a number of objectives in respect of which the quality of water is protected. The key ones at European level are general protection of the aquatic ecology, specific protection of unique and valuable habitats, protection of drinking water resources, and protection of bathing water. All these objectives must be integrated for each river basin. Only the last three -- special habitats, drinking water areas and bathing water -- apply to specific bodies of water (those supporting special wetlands; those identified for drinking water abstraction; those generally used as bathing areas). In contrast, ecological protection should apply to all waters: the central requirement of the directive is that the environment must be protected to a high level in its entirety. For this reason, a general requirement for ecological protection and a general minimum chemical standard were introduced to cover all surface waters. These are the two elements ‘good ecological status’ and ‘good chemical status’.

I. Surface water: ecological protection. Good ecological status is defined in terms of the quality of the biological community, the hydrological characteristics and the chemical characteristics. As no absolute standards for biological quality can be set which apply across the Community, because of ecological variability, the controls are specified as allowing only a slight departure from the biological community which would be expected in conditions of minimal anthropogenic impact. A set of procedures is provided for identifying that baseline for a given body of water, and establishing

[2] This section is based in large part on a text about the European Water Framework Directive published on the internet (see: http://europa.eu.int/comm/environment/water/index.html).

particular chemical or hydro-morphological standards to achieve it, together with a system for ensuring that each Member State interprets the procedure in a consistent way (to ensure comparability). The system is somewhat complicated, but this is inevitable given the extent of ecological variability, and the large number of parameters, that have to be dealt with.

II. Surface water: chemical protection. Good chemical status is defined in terms of compliance with all the quality standards established for chemical substances at European level. The Directive also provides a mechanism for renewing these standards and establishing new ones by means of a prioritisation mechanism for hazardous chemicals. This will ensure at least a minimum chemical quality, particularly in relation to very toxic substances, everywhere in the Community.

III. Surface water: other uses. As mentioned above, the other uses or objectives for which water is protected apply in specific areas, not everywhere. Therefore, the obvious way to incorporate them is to designate specific protection zones within the river basin that must meet these different objectives. The overall plan of objectives for the river basin will then require ecological and chemical protection everywhere as a minimum, but where more stringent requirements are needed for particular uses, zones will be established and stricter objectives set within them.

There is one other category of uses that does not fit into this picture: the set of uses which adversely affect the status of water but which are considered essential on their own terms -- they are overriding policy objectives. The key examples are flood protection and essential drinking water supply, and the problem is dealt with by providing derogations from the requirement to achieve good status for these cases, so long as all appropriate mitigation measures are taken. Less clear-cut cases are navigation and power generation, where the activity is open to alternative approaches (transport can be switched to land; other means of power generation can be used). Derogations are provided for those cases too, but subject to three tests: that the alternatives are technically impossible, that they are prohibitively expensive, or that they produce a worse overall environmental result.

IV. Groundwater: chemical status. The case of groundwater is somewhat different. The presumption in relation to groundwater should broadly be that it should not be polluted at all. For this reason, setting chemical quality standards may not be the best approach, as it gives the impression of an allowed level, up to which Member States are allowed to pollute. Very few such standards have been established at European level for particular issues (nitrates, pesticides and biocides), and these must always be adhered to. But for general protection we have taken another approach. This is essentially a precautionary one. It comprises a prohibition on direct discharges to groundwater, and (to cover indirect discharges) a requirement to monitor groundwater bodies so as to detect changes in chemical composition and to reverse any anthropogenically induced upward pollution trend. Taken together, these should ensure the protection of groundwater from all contamination, according to the principle of minimum anthropogenic impact.

V. Groundwater: quantitative status. Quantity is also a major issue for groundwater. Briefly, the issue can be put as follows: there is only a certain amount of recharge into a groundwater each year, and of this recharge, some is needed to support connected ecosystems (whether they be surface water bodies or terrestrial systems such as wetlands). For good management, only that portion of the overall recharge not needed by the ecology can be abstracted -- this is the sustainable resource, and the Directive limits abstraction to that quantity.

1.3.3 Institutional sustainability based on extent and coherence as criteria

Ecological sustainability depends on institutional sustainability and sustainable management as preconditions. In this book we are especially interested in the interaction between property rights and public policy and how this interaction could contribute to greater ecological sustainability. We look at institutional arrangements or regimes that have been developed through the years to manage conflicting water uses and to guide these uses in a sustainable way. Part of the regimes is that they do establish property rights and use rights over water resources, in order to clarify the ownership, but also to restrict the owner's water use by allowing others to make use of the same water resources. The possession of titles, the exclusion of uses, and the access of users are organized in this way. Another part of the regime is that supplementary policies are formulated to help these property and use rights work in the targeted directions. Studying the evolution of resource regimes, we focus on the 'extent' of a regime (i.e. the uses and use functions that are regulated by a regime and therefore belong to the regime's domain) and on the 'coherence' of a regime (i.e. the match between the regime's elements, especially the match between property rights elements and public policy elements).

The *extent* or fullness of a regime refers to the number and the restrictiveness of its rights. We could think of a continuum with on the one side extreme laissez-faire arrangements under which individual participants are free to use a resource without even the constraints imposed by some system of property and use rights (i.e. the case of 'no regime'). At the opposite extreme we find institutional arrangements featuring central planning combined with extensive structures of rules governing the actions of users. Most real-world regimes lie somewhere in between these two polar cases. Since 'extent' reflects the domain of uses and users that are regulated by the regime, it is important to explain what is meant by 'resource use'. The European Water Framework Directive defines the use of water as:

- abstraction, distribution and consumption of surface water or groundwater;
- emission of pollutants into surface water and waste water collection and treatment facilities which subsequently discharge into surface water;
- any other application of surface water or groundwater having the potential to significantly impact the status of water.

Due to the interconnection between water and land, a water resource could also be a river basin, thus including the area of land from which all surface water run-off flows through a sequence of streams, rivers and, possibly, lakes into the sea at a single river mouth, estuary or delta. Therefore, resource use may also refer to river basin use. On the basis of these notions we created for our study a list of more specific uses and use functions, which we classified as follows:

- Water for the living environment (plants and animals).
- Water for consumption and drinking water supply (domestic use).
- Water for agricultural use (irrigation or drainage).
- Water for industrial use, which means water used directly or indirectly for the production of economic goods and services (for instance, cooling as an indirect use or production of mineral water as a direct use).
- Water for hydropower production (as a particular form of economic production).
- Water resource as a medium for discharge of pollutants.
- Water resource as infrastructure for tourism, leisure, recreation, sports or medical use (e.g. bathing, swimming, skating, leisure navigation, sports fishing, wind surfing).
- Water resource as an infrastructure for commercial navigation, fishing, gravel extraction, mining, or other commercial uses.
- Water resource as an infrastructure for land use (especially use of flood plains for water storage, landscape development, urban development, etc.).

If a certain use of the water resource (e.g. fishing) is not regulated or considered by any of the regime elements, it does not belong to the extent.

Likewise, if only professional fishing is regulated, but not sports fishing, the sportsmen involved do not belong to the extent. A change in extent will often mean that more uses or use functions are incorporated. Typical of many cases is that nature (living environment) gets recognised as a use function and considered by the regime. A larger extent makes the regime more 'meaningful' for the use of the resource. But there is also a danger. If the incorporation of additional uses/users in the regime takes place by means of new, separate property rights and/or public governance aspects, this might lead to a decline of the regime's coherence. This is how simple regimes evolve into complex ones.

The concept of *coherence* refers to the degree of consistency among the elements of a resource regime. For instance, use rights frequently come into conflict with private ownership rights. Young (1983) argues that resource regimes need to be accompanied by administrative organizations and policies, especially to cope with problems of interpretation and dispute settlement. This illustrates the mutual relationship between property rights and public governance:

> "As soon as some administrative apparatus is in place, it becomes possible to think about devising techniques of social control (policy instruments) through which to guide the behaviour of those subject to a regime towards certain desired ends. Possible policy instruments are changes in bundles of exclusive rights, the promulgation of restrictive regulations, or decisions concerning individual applications for resource exploitation permits. Bundles of exclusive rights could be restructured as a means of solving dilemmas of common property in the realm of natural resources or alleviating problems of air and water pollution. Thus, we might establish licenses or entry permits relating to the use of resources. Alternatively, it would be possible to create pollution rights entitling the holder to emit specified amounts of effluent or use rights entitling the holder to make some designated use of a resource. All such rights would be transferable so that exchanges could occur and re-combinations of the rights would be perfectly feasible."

The existing property rights structure might trigger public authorities to change the property rights structure. We were able to distinguish between 'hard' and 'soft' changes of the property rights structure. Hard changes are based on some redistribution of property rights, which means that property rights are transferred from one user to another. Redistribution could take different forms, such as attribution of new property titles to new users at the cost of other users, adoption of new titles for expropriation of ownership, limitation of use rights, and transfer of private property rights to the public domain. Soft changes of the property rights structure are not based on redistribution but on compensation arrangements or on the attribution of liability. The key aim of the European Water Framework Directive to get prices right also fits into this category. Each water use will have its price in terms of affecting the status of a water resource. First, the EU wants to change the pattern of water use by making those who pollute pay the costs of the damage they cause. Therefore, the EU adheres to the *polluter pays principle*: those who pollute need to accept the financial consequences (to include externality costs). Secondly, the EU stresses the integration of the full cost of water uses into the price paid for the water, to ensure that costs are not born downstream by the rest of society and by future generations. The *full cost recovery principle* should ensure that the price charged to water users contributes to the wise use of this limited resource (also to include externality costs). Thirdly, the EU adheres to the *principle of affordability*, to ensure that basic services are provided at an affordable price (to guarantee access rights). These three principles should encourage a far more rational, sustainable use pattern. Getting the prices right could be effectuated by changing the property rights structure.

The term 'integrated regime' relates to the term 'integral water management', which is high on the agenda of the European Union and its member states. In the European Water Framework Directive 'integrated management' is not described in social science terminology, but in more practical terms. In the Euwareness project and hence in this book 'integrated management' is given a more defined and elaborated meaning than usual. The term 'integration' is used as a label for the regime changes that might improve both the extent and the coherence, and thus the sustainability record of water resource regimes.

We understand regime coherence as the degree of consistency between the property rights structure and the governance structure of a resource regime. We are also interested in the internal coherence of the property rights structure itself and in the internal coherence of the governance structure. Therefore, we distinguish between the following change variables:
1. Extent of a regime.
2. Internal coherence of the property rights structure of a regime.
3. Internal coherence of the governance structure of a regime.
4. External coherence between the property rights structure and the governance structure of a regime.

The criteria for 'integral water management' as provided by the European Water Framework Directive mostly refer -- in our terms -- to a combination of an increased extent and more internal coherence of the governance structure of a regime. Only the criterion of 'getting the prices right' appeals to an improvement of the external coherence between the property rights structure and the governance structure of a regime. It is apparent, therefore, that options for the improvement of the external coherence -- in particular the options we suggest for the redistribution of property rights -- are not considered by the European Water Framework Directive.

In the next chapter we explain that the governance structure of a regime can be analysed along five dimensions or elements. Therefore, the internal coherence of the governance structure of a regime can be parsed into:

- Coherence of levels and scales of governance;
- Coherence between actors in the policy network;
- Coherence of problem perception and objectives;
- Coherence of strategy and instruments;
- Coherence of responsibilities and resources for implementation.

The options for improving integral water management, as advocated by the European Water Framework Directive, can be linked to these five forms of coherence:

- Water management based on river basins: administrative co-ordination at the level of a river basin as a whole ("*the best model for a single system of water management is management by river basin - - the natural and hydrological unit -- instead of according to administrative or political boundaries*" – EU, 2000). This element supports the form of coherence focusing on levels and scales of governance.
- Getting users involved: involvement of all actors having an interest in water services ("*increasing public participation and balancing interest of various groups*"– EU, 2000). This element supports the form of coherence focusing on actors in policy networks.
- Expanding the scope: development of a water vision for a river basin ("*co-ordination of objectives – good status for all waters by 2010; the objectives for a river basin must be set out in a river basin management plan, based on analysis of the river basin characteristics, a review of the impact of human activity on the status of waters in the basin, estimation of the effect of existing legislation and the remaining 'gap' to meeting these objectives*" – EU, 2000). This element supports the form of coherence focusing on problem perception and objectives.
- Streamlining legislation ("*the framework directive will take over operative provisions of several water directives*" – EU, 2000) and getting the prices right by full cost recovery pricing ("*to ensure that the price charged to water users integrates the true costs*" -- EU, 2000). This

element supports the form of coherence focusing on strategy and instruments.
- Co-ordination of implementation: co-ordination of the application of measures for a river basin ("*analysing whether existing legislation solves the problem once and for all, and if it does not, identifying why and designing whatever additional measures are needed to satisfy all the objectives established*" – EU, 2000). This element supports the form of coherence focusing on responsibilities and resources for implementation.

An example of an inconsistency between the elements of public governance is when a new problem perspective is accepted, but no new targets are formulated for that newly recognised problem, or it is not recognised that the new targets are contradictory to the existing ones.[3] This would thereby create the risk that 'the left hand knows not what the right hand doeth'. These are examples of a mismatch within an element. When the new objective is not followed by instruments to attain it, that is a mismatch between elements of governance. An example of a misfit in the property rights system is when new users are granted use rights without recognising that this may harm existing use rights[4], for instance when water scooters are allowed in a lake where a sports fishing association holds an exclusive fishing right. An example of lack of coherence between property rights and public governance occurs when policy instruments address other actors than those that hold relevant use rights.

1.4 The case studies on water basins in this book

As explained before above, our study was based on screenings of the national regime evolution in six European countries, followed by two case studies in each country at the water basin level. The case studies focused more or less on the period 1970-2000 in order to gain an understanding of the conditions that have generated regime changes from complexity towards a more integrated status. This book presents the comparison of the case studies. The case studies consist of a detailed examination of the development of more integrative regimes. With the help of this information we have tested hypotheses on the generation and effects of such regimes in

[3] In the French Audomarais case (see chapter 5) nature protection was added, while the groundwater extraction only reckoned with industrial use and drinking water production and resources for that, but not with the needs of the natural area.

[4] In the French Audomarais case (see chapter 5) the water level in the canal was raised by the State company holding the management rights to allow bigger boats, but without reckoning on the rights of the users of the water for market gardening.

practice. The cases were water-basin-based and of intermediate scale (mostly tributary basins).[5] Cases of regimes studied addressed issues of surface water, groundwater, wetlands or combinations of these. The reason for carrying out the project at a European level is that the involvement of six European countries and twelve case studies with a widespread geographical and institutional variety create an interesting learning potential. The twelve cases display a wide variety of rivalries, including most of the common ones in the European Union at this level. They are similar, however, in their approximate scales. The following cases were selected[6]:

Netherlands

Case 1 -- The *IJsselmeer* is a freshwater lake and former Inland Sea in the heart of the country. It covers about 2000 km^2, surrounded by about 600 km of shore. Rivalries in use concern fishery (a homogeneous use rivalry), between gas drilling and the protection of water quality for nature and drinking water production, the rather Dutch issue of land reclamation vs. open water, and between nature and recreation.

Case 2 -- The *Regge River Basin* is a tributary river in the Province of Overijssel, joining the Vecht River, which ultimately flows into the IJsselmeer. It is a small rain river with only 20 meters of height difference in a 900 km^2 basin. The main rivalries are between protection against flooding (mainly for agriculture) and nature and landscape, and between pollution and nature.

Belgium

Case 1 -- The *Vesdre River Basin* is a tributary river in Wallonia. Its basin is 710 km^2 and it contains two rivers, de Vesdre (71 km long) and the Hoëgne (29 km long). It is an independent basin, since all rivers have their source inside the basin. Besides quality aspects, quantity aspects are also relevant in this case. Since it is a 'wet' case, here the latter means the risk of flooding.

5 According to the European Water Framework Directive a river basin means the area of land from which all surface water run-off flows through a sequence of streams, rivers and, possibly, lakes into the sea via a single river mouth, estuary or delta. The directive also distinguishes sub-basins. A sub-basin means the area of land from which all surface run-off flows through a series of streams, rivers, and, possibly, lakes to a particular point in a water course (normally a lake or a river confluence). In the Euwareness project we have concentrated on sub-basins as unit of analysis for the cases.

6 Of course, far more extensive information about these cases is provided in the twelve Euwareness case study reports, which form a separate and essential part of the reporting on the Euwareness research.

Case 2 -- The *Dender River Basin* is also a tributary river, but now in Flanders. Its basin is very similar (708 km^2) in size. In fact it is larger (1384 km^2), but half of it lies in Wallonia. The Dender flows only in its lower part into the basin. Both flooding and quality aspects are relevant. There is no drinking water production in the basin.

France
Case 1 -- The *Audemarois Basin*, including the Aa River Basin and the Audemarois Marsh as its spreading area. The river Aa is 40 km long and is the main river of the region. The borders of the Audomarois Regional Park, encompassing some 550 km^2, best define the area. There are various rival uses at stake: agriculture and market gardening vs. navigation and nature, industrial pollution and nature, and nature protection vs. extraction for drinking water and industry.

Case 2 -- The *Sèvre Nantaise River Basin* is rather large, 2,493 km^2, but not very densely populated (290,000 in 1999). There are several smaller rivers in the area, but the Sèvre Nantaise itself is 135 km long. It is in turn a tributary of the Loire. The rivalries consist of agricultural drainage and irrigation vs. nature (because of floods and low water periods), and pollution from different sources and nature.

Spain
Case 1 -- The *Matarraña River Basin*, a tributary of the river Ebro located in the northeast of Spain. The Matarraña is a tributary of the Ebro river and has an extension of 97 km. The basin is 1,727 km^2 large. In this case irrigation -- using 90% of the available water flow! -- is rivalling uses for population supply, tourism and environmental protection.

Case 2 -- The *Mula River Basin*, where irrigation by itself is already pushing to the limits, but also rivals uses for population supply, tourism and environmental protection. The Mula is a tributary of the Segura and has an extension of about 25 km. Its basin is 695 km^2.

Italy
Case 1 -- The *Idro Lake*, a natural basin, artificially regulated, and the *Chiese River*, that flows across the regions Trentino and Lombardia, generating the Idro Lake and finally flowing down the Sabbia Valley into the River Oglio. The Chiese flows for 148 km. The whole water basin covers 934 km^2. The Idro Lake is a natural lake, artificially regulated. Its surface is 11 km^2; its total catchment area covers 617 km^2. The main rival uses are agriculture, hydropower, tourism, nature and protection from floods, soil

erosion and land slip. The rivalry focuses on the maximum variation of the water level of the lake and a minimum river flow.

Case 2 -- The *Marecchia-Conca Basin*, where the main rivalry is between water for drinking (residents, but also tourists) and irrigation for agriculture. But there are also some other rivalries and institutional conflicts. The basin covers 1347 km^2. The water basin includes the Marecchia cone, the biggest hydrological sink of the region.

Switzerland

Case 1 -- The *Maggia Valley*, which is a mountain river basin situated in Ticino, the southern part of Switzerland, south of the Swiss Alps. The river flows into Lake Maggiore, a large part of which is situated on Italian territory. The water basin is about 930 km^2 big. Main problems were with hydroelectric production, quarrying and flood protection, rivalling with uses like recreation, drinking water production and nature and landscape.

Case 2 -- *Lake Baldegg and Lake Hallwil*, where the problems stem from wastewater from settlements, diffuse pollution by agriculture and the need to protect the lake shores. The first lake is 5 km^2 in size, mainly supplied by river Ron. From Lake Baldegg flows river Aabach that feeds the second lake of 10 km^2. The water basin of the two lakes is relatively small, some 138 km^2.

Instead of just including the case studies as separate chapters in this book, we have chosen to write country-based chapters, in which the two case studies are summarized, linked together and linked with the national regime screening in the other book (Kissling-Näf and Kuks, 2004). The chapters concentrate on the factors that explain the success/failure of the attempts at regime integration that we generally find in all the countries. The chapters have broadly the same structure (an introductory paragraph describing the most recent attempts to integration, the interpretation of the case studies, a conclusion), but already in the title and later in the conclusion they focus on the most important feature that the authors believe characterizes the country.

In chapter 2 we explain the theoretical backgrounds of the case studies, the hypotheses underlying the research, and the methodological backgrounds and choices that have been made.

Chapter 3 tells the story of the *Netherlands*, a delta area for three European river basins. The need to protect the land from high water from rivers and sea, and the tradition of artificially draining low-lying areas, has given the country a complex hydraulic infrastructure. In the 1960s and the 1970s the traditional water engineering approach started to come under fire, which resulted in the adoption of rival water values and the greening of water engineering. From 1985 on, the ecological aspects of water systems have been incorporated in water management, shifting the regime from a complex to an integrated status. Since the early 1990s the country has become involved in a paradigm shift in its approach to flood protection, trying to store water in retention areas, while maintaining but not expanding the country's infrastructure of dyke fortifications. Especially in flood plains, new rivalries are evolving between water use and other land uses. This requires not only integrated water management, but also integration between water and land use management.

Chapter 4 is about *Belgium*, a federal state with two regions (Flanders and Wallonia) which are rather autonomous in their water management since 1993, when the Belgian state became fully federal. However, the complicated process of federalisation delayed -- compared to other European states -- an effective approach to water problems. The Belgian case is interesting in that it shows two regions with different types of public domains. Although both component states show an expanding public domain, in Flanders the public domain is based on private ownership by the state, while in Wallonia the regulation of use rights on the basis of public law is a way to get public control. In Flanders water management procedures are more centralised and more subject to central planning, while in Wallonia the local level is much more autonomous and water management is characterised by bottom-up decision making. Neither regional system is sufficiently integrated. The Flemish system lacks participation by local actors, while the Walloon system is poor at planning.

Chapter 5 shows that water basin management in *France* had an early start. As early as 1964 the country adopted water legislation which created water agencies at basin scale, recognising regional variation and the need for specific solutions. Almost thirty years later, in 1992, France adopted another important water act, based on improved integrative thinking, promoting planning, creating local institutions, better addressing environmental issues, for instance by introducing a better application of cost recovery and the polluter pays principle. The French regime is strong in taking account of regional differences by means of innovations in administrative organisation and planning. However, water management has mostly been governance driven; the property rights structure remains hardly affected. Until now, this is a weak point in achieving a further integration of French water management.

Chapter 6 describes the situation in *Spain*, where water scarcity dominates the agenda water management. The strategies for dealing with water scarcity seem quite diverse and broadly contrast between those in favour of a supply approach and those in favour of a demand approach.

While policies at the national level seem to favour the former approach, which consists of the construction of large hydraulic infrastructures transferring water between river basins, there are a few experiences in which this traditional approach has been replaced by a new one in which rationality of water use is a guiding principle. The two cases analysed in this chapter constitute two experiences with demand management. Both cases show regime changes that are property rights driven, but they vary in the strategy that has been followed. One case shows a bottom-up process opening doors to a multi-actor negotiation pattern. In the other case regime change is brought about by a top-down process in which central actors have initiated a change of the property rights structure.

Chapter 7 explains that water management in *Italy* is incoherent due to the fact that it struggles with three competing integration principles, all of which have been introduced in the last 15 years. These principles have a different definition of the problem, they assume different constellations of actor involved, and they have different implementation problems. The first integration principle was adopted in 1989. It promotes integration at the scale of river basins, mostly concerned with the quantitative dimension, and managed by a network of water basin authorities. The second principle, adopted in 1994, advocates integration at the scale of the optimum area for water supply and purification, mostly concerned with the establishment of an integrated water service, and therefore with water as a commodity, and managed by a network in which the regional governments and local authorities play a major role. The third principle, adopted in 1999, introduces integration at the scale of the water body, mostly concerned with the qualitative dimension, and managed by a network in which the regional governments and the environmental administration seem to be the key actors. The implementation structure for these three principles is both too complicated and too simple. Especially local authorities have too many roles to play at the same time, which results in an unstable regime, because they have to choose which role to play.

Chapter 8 deals with *Switzerland*, often typified as the 'water tower' of Europe due to its high precipitation and relatively huge freshwater stock. The country has a (con)federalist system where cantons often function as laboratories for national solutions. The heterogeneous geographical situation leads to very different solutions to water problems. The strong position of the cantons gives room for highly variant resource regimes at regional level, in which the property rights structure appears to be stronger than in most other European countries (except for Spain). At the level of the confederation Switzerland is somehow silently and gradually adapting to European standards and directives, due to close economic and trade relations with the EU. However, the country is not easily adopting the river basin approach as advocated by the EU, since its water regime is fragmented and organised along rivalry lines. Three important institutional arrangements for water management are identified, which are separated along three traditional issues: flood protection (recognised by 19th century legislation); utilisation of water, mainly for hydropower (recognised by early 20th century legislation); and water protection (recognised by water quality legislation in the 1950s, and strengthened by additional legislation in 1975 and 1991which added a quantitative dimension). These three separate policy communities appear to be very persistent and only tend to open up under very strong pressure.

In Chapter 9 the cases that were studied in this research are used for a comparative analysis and some implications of these results are discussed. Here we will return to the starting point in this introductory chapter: what insights could be helpful for the implementation of the European Water Framework Directive?

REFERENCES

Bressers, Hans; Kuks, Stefan (2001) Governance patronen als verbreding van het beleidsbegrip, in: *Beleidswetenschap*, Vol. 15, No. 1, pp. 76-103. Also to be published as: What does governance mean? From concept to practice, in: Hans Th. A. Bressers and Walter A. Rosenbaum (eds.) *Achieving sustainable development: The challenge of governance across social scales*, New York-Westpoint-London: Praeger (2004).

Bromley, Daniel W. (1991) *Environment and Economy. Property Rights and Public Policy.* Oxford UK /Cambridge USA: Blackwell.

EU (European Union) (2000) *European Water Framework Directive.* Published on the internet (http://europa.eu.int/comm/environment/water/index.html).

Kissling-Näf, Ingrid; Varone, Frédéric (2000a) Historical Analysis of Institutional Resource Regimes in Switzerland. A Comparison of the Cases of Forest, Water, Soil, Air and Landscape. 8th Biennial Conference of the International Association for the Study of Commen Property (IASCP) “Crafting Sustainable Commons in the New Millennium”, Bloomington.

Kissling-Näf, Ingrid; Varone, Frédéric (2000b) *Institutionen für eine nachhaltige Ressourcennutzung. Innovative Steuerungsansätze.* Chur, Zürich: Verlag Rüegger.

Kissling-Näf, Ingrid; Kuks, Stefan (eds.) (2004) *The Evolution of National Water Regimes in Europe. Transitions in Water Rights and Water Policies towards Sustainability.* Dordrecht-Boston-London: Kluwer Academic Publishers.

Knoepfel, Peter; Kissling-Näf, Ingrid; Varone, Frédéric (2000) *Broadening Property Rights Theory by Policy Analysis.* Cahier de l'IDHEAP, Chavane-près-Renens.

Knoepfel, Peter; Kissling-Näf, Ingrid; Varone, Fred (2001) *Institutionelle Regime für natürliche Ressourcen. Boden, Wasser und Wald im Vergleich.* Basel: Helbing & Lichtenhahn.

Ostrom, Elinor (1990) *Governing the Commons, The evolution of institutions for collective action.* Cambridge: University Press.

Young, Oran R. (1983) *Resource Regimes. Natural Resources and Social Institutions.* Berkeley-Los Angeles-London: University of California Press.

Chapter 2

Institutional Resource Regimes and Sustainability

Theoretical backgrounds and hypotheses

Hans Bressers, Doris Fuchs and Stefan Kuks
University of Twente (Enschede-Netherlands)

2.1 Introduction

This chapter does not attempt to present the voluminous world literature on the topic of institutional regime analysis. Only a small selection of this literature will be dealt with. Our main purpose here is to present the theoretical notions that we have chosen or developed -- standing on the shoulders of many others -- and used in this particular research project. Furthermore, we formulate our hypotheses about regime shift towards integration and the implications of institutional resource regimes for sustainability.

In this book we interpret 'regimes' as institutional resource regimes, comprising a public governance component and a property rights component. The combination of those components can be more or less integrated and influences the sustainability of the use of the given natural resource. In turn, these regimes, or rather their property rights and governance components, are influenced by external change agents, which leads to regime change. Figure 2.1 illustrates these dynamics[1] as will be investigated in the case study comparison in Chapter 9. As the figure shows there are three groups of variables. These are linked by the central relationships in the research questions:

[1] We acknowledge the existence of several other possible feedback relations, but they are not presented in the graph since they receive less attention in our discussion.

Hans Bressers and Stefan Kuks (eds.), Integrated governance and water basin management. Conditions for regime change towards sustainability, 23-58.

1. How far do more integrated water resource regimes lead to more sustainable resource use?
2. What change agents and conditions cause shifts towards more integrated regimes?

The integration of the 'institutional resource regime' is the central variable. Question 1 should show the results of such integration. Question 2 should provide explanations for it. Therefore, we start by explaining the regime components themselves. In section 3 we deal with the factors that might contribute to the integration of resource regimes. Section 4 deals with the effects on sustainable resource use of regimes. Section 5 summarises the hypotheses that are tested in this study. Section 6 goes into the methodology of the study and the comparison.

Figure 2.1*: Research model*

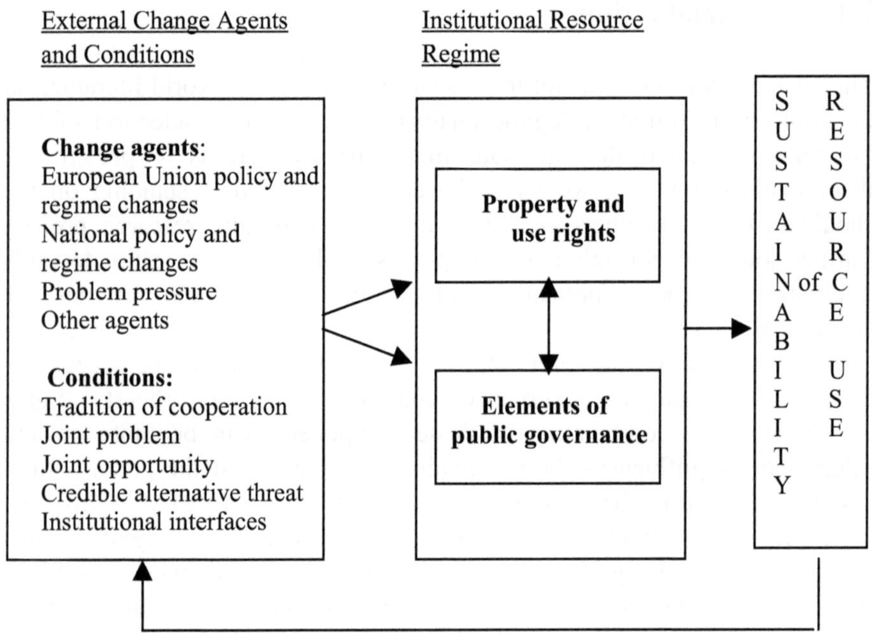

2.2 The public governance and property rights components

2.2.1 Public governance

The debate on recent modes of regime analysis is often conducted under the flag of 'governance'. Yet 'governance' is a broad and often confusing term because it is used in many different ways (Peters and Pierre, 1998; Lynn et al., 2000a, 2000b: 234; O'Toole, 2000: 276). There are almost as many ideas about governance as there are researchers in the field (Björk and Johansson, 2000: 1).

Rhodes identifies the four categories of literature that are most relevant to our concern with defining 'public governance' as the minimal state, new public management, socio-cybernetic systems theory, and self-regulating networks (Rhodes, 1997; Björk and Johansson, 2000: 3-7). The factor common to all these categories is that they search for forms of co-ordination that do not fit comfortably into the 'market'–'hierarchy' distinction (compare Arentsen and Künneke, 1996; Rosenau, 2000: 11-16) once common to governance theory. At the beginning of the 1980s there was talk of 'overloaded polities' in which 'regime stability is threatened by the related problems of governmental effectiveness and popular consent.' (Rose, 1980). The reaction to this was advocacy of 'less state and more market'. Thinking in terms of governance is also an answer to the 'erosion of the state' literature during the 1980s. Rhodes (1997: 19) writes: "The process of hollowing out…is central to understand the shift from a unitary state to a differentiated polity."

Each of these approaches is to some degree built up from empirical observation but, more importantly, they rest on a normative vision. This normative bias almost always implies that a more limited role or presence of government authorities delivers better governance. This bias is weakest in the 'socio-cybernetic approach' (e.g. Kooiman, 1993) and most obvious in the idea of the minimal state. The normative debate described above is not our primary interest, even though, especially for the policy sectors that are relevant to sustainable development, many authors stress that public governance is very important since the main beneficiaries (people in faraway countries and of future generations) are not capable of defending themselves in national arenas. Implicitly, our position is that 'governance without government' (Peters and Pierre, 1998: 223) is not a goal in itself.

Some authors are looking mainly for a more or less stable institutional arrangement in addition to the market, in which communities of private actors are able, without 'outside intervention', to promote their collective interests. In contrast, our interest does lie in such 'outside interventions', but against the background and in combination with 'private' rules-in-use, customs and traditions, property rights (compare Young, 1994: ix and 163). Although Young looks for the substance of governance in institutional arrangements, March and Olson (1995: 11) emphasize ideas of governance opposing the institutional perspective, as a consequence of the 'exchange' perspective. For them, 'the core artistry of governance is winning coalitions and policies'. Clearly, we do not agree with either standpoint. In this study, property and use rights are studied as a separate part of the institutional resource regime (see next subsection).

Multi-actor networks
Many changes in government co-ordination strategies *vis à vis* the public sector are related to the growing recognition that government alone does not determine societal developments; these are shaped by many actors, public and private. Within such networks of actors, government can adopt a position that is more or less central and dominant. This change in view represents a shift in accent from government policy, or 'government', to 'governance'. In this research, we speak mostly of 'public governance' to emphasise that this part of the institutional regime addresses mostly the public sphere. The 'governance' pattern consists of all of the consequences of the interplay among all the actors involved (Kooiman, 1993: 258) in interventions to promote societal goals and the production of collective goods.

Multi-level co-ordination and multi-faceted problems
In addition to the multi-actor aspect, there is a growing recognition that sectors in society are not governed on one level, or on a number of separate levels, but through interaction between these levels. These levels often but not necessarily, reflect the various tiers of government; other powerful actors may provide direction at a certain level where no government authority is active. In the same way, actors may also operate on more than one level. Thus, there is a growing recognition that policy problems often contain various interacting levels (e.g., problems related to sustainable development). This reality reflects what has been identified as 'multi-level governance'. The term 'multi-level' has become the most common prefix attached to 'governance'. It relates not only to its multiple nature but also, and in particular, to the mutual interdependence between levels (Smith, 1997). Blomquist and Schlager (1999: 7, 39-43) also emphasize the relation between the many facets of environmental problems and the horizontal and vertical co-ordination this requires, as does Rosenau (2000: 10-11).

Multi-instrument strategies and multi-resource-based implementation
O'Toole (2000: 276-279) treats governance in the context of studies of the implementation of policy strategies. He states that governance is 'difficult to denote with precision', adding to the multi-level and multi-actor aspects 'the multivariate character of policy action'. He refers to Milward and Provan (1999: 3), who state: 'The essence of governance is its focus on governing mechanisms -- grants, contracts, agreements -- that do not rest solely on the authority and sanctions of government.' He also points to the work of Lynn et al. (2000a and 2000b), who approach governance from the public management perspective. Although they set themselves the task of developing a broad and comprehensive model of governance, their background is clearly present in their thinking. They begin by noting that policy programs are implemented in a web of many diverse actors, an assumption that differentiates their approach from the rest of the literature. As a consequence, the model of governance they develop concentrates not only on the objectives (including output indicators) and instruments ('treatment') of policy, but also the resources and organisation of implementation. The model differs from the usual reviews mainly because it clearly shows that these aspects of organisation and resources can take a wide variety of forms and have a multi-functional character (pp. 257-258). Peters and Pierre (1998: 226-227), besides the emphasis on networks, also consider the 'blending of public and private resources' and 'the use of multiple instruments' to be features of the governance concept.

A model of governance in five elements
Various contemporary approaches in policy science focus on changes in government policy and other modes of co-ordination, especially when making comparisons (Sabatier and Jenkins-Smith, 1993, 1999; Baumgartner and Jones, 1993). In addition to a focus on long-term changes in policy (diachronic study), there are also comparative studies of policies in a certain sector in different countries (synchronic study). For both these purposes in this research, it is necessary to picture what we understand to be the 'elements of governance' that deserve attention.

Partly based on the previous discussion and partly on an examination of contemporary policy science approaches (Dryzek, 1987, 1997; Fischer, 1995; Fischer and Forrester, 1993; Kingdon, 1995 (1984); Rhodes, 2000; Sabatier and Jenkins-Smith, 1999; Scharpf, 1997; Schön and Rein, 1994; Schwarz and Thompsom, 1990;Thompsom, Ellis and Wildavsky, 1990; Zahariadis, 1999) we arrive at the following five elements of public governance (Bressers and Kuks, 2001, 2002). The questions below the labels are not posed here to be answered in the study, but to operationalise what is meant by these elements of governance.

(1) *Levels and scales of governance*
Where? – Multi-level
Which levels of governance dominate policy and the debate on conducting policy, and in which relations? What is the relation with the administrative levels of government? Who decides or influences such issues? How is the interaction between the various administrative levels arranged?

(2) *Actors in the policy network*
Who? – Multi-actor
How open is the policy arena in theory and practice, and to whom? Who is actually involved and with what exactly? What is their position? What is the accepted role for government? Which actors have relevant ownership and use rights or are stakeholders in some other capacity (including policy-implementing organisations)? What is the structural inclination to co-operate among actors in the network? Are their 'coalitions' formed across social positions? Are there actors among them who operate as process brokers or 'policy entrepreneurs'? What is the position of the general public versus experts versus politicians versus implementers?

(3) *Problem perception and policy objectives*
What and why? – Multi-faceted
What are the dominant maps of reality? What is seen as a problem and how serious is this considered to be? What do people see as the causes of the problem? Is the problem considered to be a problem for individuals or a problem for society at large? Why are the previous questions seen that way: what values and other preferences are considered to be at stake? Which functions are allocated to the sector? Is the problem seen as a relatively new and challenging topic or as a 'management' topic without much political 'salience'? To what degree is uncertainty accepted? Where are the recognised points of intervention? What relations with other policy fields are recognised as co-ordination topics? Which policy objectives are accepted? What are levels to which policy makers aspire (ambition) in absolute terms (level of standards) and relative terms (required changes in society)?

(4) *Strategy and instruments*
How? – Multi-instrument
Which instruments belong to the policy strategy? What are the characteristics of these instruments? What are the target groups of the policy and what is the timing of its application? How much flexibility do the instruments provide? To what extent are multiple and indirect routes to action used? Are changes in the ownership and use rights within the sector anticipated? To what extent do they provide incentives to 'learn'? What requirements do they place on the availability of resources for implementation? How are the costs and benefits of the policy distributed?

(5) *Responsibilities and resources for implementation*
With what? – Multi-resource-based
Which actors (including government organisations) are responsible for implementing the policy? What is the repertoire of standard reactions to challenges known to these organisations? What authority and other resources are made available to these actors by the policy? With what restrictions?

So the public governance component developed in the theoretical framework consists of five elements. These five elements provide answers to the five central questions of governance: Where? Who? What? How? and With what? Furthermore, a characteristic feature of modern 'governance' systems is that they have many aspects. They are multi-level, multi-actor, multi-faceted, multi-instrument and multi-resource-based. The assumed relationships between these five elements are based on the basic principle that the elements of public governance each form the context of the other elements and that they will tend to adjust to each other.

2.2.2 Property and use rights

Property rights arrangements are the second important component of an institutional resource regime. In the context of resource research, the property rights approach is particularly worthy of mention (Bromley, 1989, 1990, 1991; Burns and Dietz, 1996; Feeny et al., 1990; Schlager and Ostrom, 1992; Libecap, 1993; Devlin and Grafton, 1998), specifically the common-pool resources theory (Ostrom 1990; 1992a; 1992b; 1994; 1997; Ostrom et al., 1994). According to Ostrom, property and use rights exercise a decisive influence on the use of natural resources in that they determine who has access to the resource and when and in what form it can be used. The prevailing system of property rights assignments in the community is, in effect, the set of economic and social relations defining the position of interacting individuals with respect to the utilisation of scarce resources (Pejovich, 1975: 40).

Property rights delineate rights of ownership in an asset, which generally include the rights to use and consume the asset, to exclude others from the use of the asset, to change its form and substance, to obtain income from it, and to transfer these rights either in their entirety through sale or partially/temporarily, for instance through rental (Barzel, 1989; Furubotn and Pejovich, 1975; Kasper and Streit, 1998). While property rights may be exclusive, they are generally not unrestricted.[2] Governments, for example, often impose regulations limiting the owners' options in terms of how they can use their resource.

Property rights should be conceived of as bundles of rights. With respect to environmental resources, for instance, property rights exist and frequently differ for the stock of a resource and the produced yield or the goods and services derived from a resource. 'Ownership of the resource' would thus pertain to a specific bundle of rights the owner holds with respect to the resource. The owner may, for example, hold the right to farm the land, but not to kill rare species on the land.

There are limits to the bundling and unbundling of property rights from a legal perspective, of course, as some contractual agreements on specific

[2] This fact is important to remember in the context of debates on the environmentally desirable property regime. In this debate, private property rights are often treated as absolutes, which in reality they rarely are. Rather than having to choose between private property regimes, common property regimes, and state ownership, the imposition of some constraints on private property is often a reasonable alternative.

bundles of rights might be viewed as unconstitutional or immoral.[3] Furthermore, property owners may not be able to unbundle a particular right, often a negative right, i.e. responsibility associated with their property ownership. In general, however, the perspective on property rights as bundles of rights cannot easily be rejected.

Notions about which specific rights generally go with the ownership of a good or asset and the extent to which it is possible to unbundle specific rights differ across time and culture. Thus, it is much more common to differentiate between specific rights to attributes of a good or asset in the Common Law tradition than in German law, for instance. This is somewhat ironic, since the Common Law tradition happens to be based on old Germanic legal frameworks (in which this unbundling was possible). In contrast, German law today is essentially based on the Roman concept of property rights, to which the countries on the continent reverted in the context of the 'creation' of the concept of territorial sovereignty associated with the Westphalian peace in the 17th century. Under Roman law, unbundling was generally much more limited, especially with respect to real estate. Thus, the owner of real estate owned everything below the area of land as well as the space above the area of land. In consequence, owning apartments, for instance, was not possible under Roman law. Despite this difference in traditions, notions of bundling and unbundling are not set down in stone, of course. Thus, unbundling has become much more common on the European continent these days, while in the United States property owners are fighting against the right of government to interfere with specific rights associated with their property.

With respect to natural resources, the focus on 'bundles of rights' highlights that property rights to multiple attributes of a resource can exist and be held by different individuals or groups. Different 'property regimes' are likely to exist with respect to the attributes of many environmental resources. Property rights and regimes for such a resource thus tend to form a complex structure with several layers and dimensions.

In the context of this project, the 'coherence' of this layering of property rights is of particular importance, since the project aims to emphasise heterogeneous demands for water resources. In contrast, much of the literature to date, such as Ostrom's early irrigation studies, has focused on homogenous uses. One way to characterise the difference between the two situations may be to distinguish between competing and conflicting property rights. In the case of scarcity in a homogenous use situation, property rights are competing with each other, while in the case of scarcity in a

3 Such limits on unbundling can also exist in terms of the feasibility of the separation of property rights to different attributes to the resource.

heterogeneous use situation, property rights are also conflicting with each other. The objective of policy intervention in the context of an institutional resource regime, then, can be to lead to a co-ordination and harmonisation of rights to different attributes of the resource and pursue sustainable management through a reduction in conflict between these rights.

As Stubblebine (1975) argues, the definition of property rights becomes necessary as soon as two individuals share a living environment. With Friday's arrival on Crusoe's island: "Some set of property rights must, and will, be created to condition the relationship between these two individuals - whether that set be characterised as capitalistic, socialistic, or something else" (p. 13). This definition of rights will take place, whether or not the institution of 'property rights' has been intentionally created by the community or by an outside authority and enforcer.

In combination with social norms, available technologies, and resource conditions, property rights yield collective outcomes in social settings (Young, 1994). They determine the allocation and use of resources, composition of output, and distribution of income which result from interacting decision makers: "Such institutions critically affect decision making regarding resource use and, hence, affect economic behaviour and performance. By allocating decision-making authority, they also determine who are the economic actors in a system and define the distribution of wealth in a society" (Libecap, 1993: 1). In an increasingly complex world with interactions between decision-makers across space and time, the definition of relationships through property rights provides the foundation for environmental and economic activity.

The institution of property rights can take a variety of forms. Property institutions "range from formal arrangements, including constitutional provisions, status, and judicial rulings, to informal conventions and customs regarding the allocations and use of property" (ibid.). In contrast to traditional societies with dense social networks and accordingly low transaction costs, the interactions in today's Western societies require "elaborate institutional structures that constrain participants and minimise transaction costs" in the form of well-specified and well-enforced property rights (North, 1989: 1320f).

While we tend to think of property rights as long-term, stable relations of ownership, such rights are never constant. From an economic perspective, property rights change as individuals desire to adjust to changed economic, political, and social conditions. They thus create demand for a change in property rights until they are satisfied and have reached a new 'equilibrium' position in rights over resources. Property rights to resources are determined by the interaction of supply and demand in dynamic sequences. "Every individual seeks those property rights modifications which he believes will improve his welfare" (Stubblebine, 1975: 15). Property rights, then, are created or changed in response to economic forces, as opportunities to gain arise (North, 1989: 1324; Feeny, 1988: 273; Ensminger and Rutten, 1991).

In contrast to the economic perspective, legal and political philosophy would emphasise the tendency of property rights to resist and work against change. Thus, one can argue that individuals tend to perceive property rights as rather permanent, and do not constantly perform cost-benefit calculations to determine the current best allocation of property rights for them. Furthermore, they might value 'property ownership' independently of its direct economic return, because of personal values or status considerations, for instance. In addition, individuals are also guided in their actions by norms and habits, which tend to work against constant and rapid change on the basis of rational calculations.

In order to assess the role of property rights and especially the implications of property rights arrangements for sustainability later on, two fundamental propositions need to be accepted. First, property rights need to be considered as *bundles of rights* (Barzel, 1989). Rather than viewing property ownership as the ownership of an entire good or resource, an analysis needs to scrutinise those attributes of a resource to which the 'property owner' actually holds rights. Thus, a land owner might own the right to use his property for farming or cattle grazing, but nowadays usually does not own the right to pollute with toxic substances the ground water or any river on or next to his property.

Secondly, the sustainability of a given distribution of property rights depends on the *de facto* property rights of the appropriators from the resource more than their *de jure* property rights (Schlager and Ostrom, 1992). Thus, an analysis needs to consider which specific property rights actually can and are being used, rather than which property rights are legally held. Of course, legal title might give some insight into possible future developments. At any given point in time, however, the de facto distribution of property rights is extremely important.

The literature on environmental resource management generally differentiates between different property regimes, i.e. property arrangements characterised by different combinations of property rights, in terms of ownership, access, and withdrawal regulations. In this book we distinguish between the 'proper' property rights (including not only the property titles, but also management rights and exclusion rights), access and use rights and disposal rights (a special kind of use right, but special in the sense that disposal directly affects water quality and its protection).

2.3 Change toward integrated resource regimes

In general, we expect the elements of public governance (and the regime in general, i.e. including property rights) to exert a stabilising influence on each other. This stabilising influence occurs through processes of mutual adaptation of values, cognitions and resources. Thus, while changes in the elements of the governance pattern can be caused by changes in other elements, ultimately these changes must have external sources affecting one or more elements from the outside. Mutual adaptation mechanisms that, without external 'disturbances', have a stabilising influence then become the mechanisms by which substantial changes in one of the elements are followed by responding changes in other elements, resulting in complete regime changes.

Sources of change

In principle, external change agents can enter the scene through all of the elements that are discerned in the regime. There is a difference, though. Property rights might be conceived as somewhat more stable and less oriented towards invoking change than the elements of public governance. That means that, although property rights may act as a powerful context for developments in public governance, changing governance patterns is not their subject per se. On the other hand, interventions from the governance side often have the specific and deliberate intent to change property rights.

Change agents for property rights may develop from three sources: (a) changes in the general cultural and judicial conceptions of property and its meaning in terms of specific bundles of rights, stemming from general policy institutions and policy processes as a context, (b) economic changes, some with technological, demographic, cultural, or spatial developments as drivers, that change the value of certain uses and of the resource itself, and (c) specific and often deliberate influences from the governance pattern on water resource management to adjust property rights as a means of promoting policy goals.

External change agents for the governance pattern stem from changes in political institutions, in the general policy process or policy processes in related fields, the spectrum of technological, demographic, and cultural developments mentioned above, as well as feedback from the actual problem situation. As examples, related to the subject of this project, below some specific and general external sources of change are linked to the five elements of public governance (cf. Bressers and O'Toole, 1995):

Levels and scales of governance
- Rise of the European Union
- Tendency to multi-level governance

Actors in the policy network
- Rise of environmental and nature organisations
- Tendency to multi-actor governance: increased number of actors involved in relevant networks

Problem perception and objectives
- Rise of environmental degradation information
- Tendency to incorporate multiple perspectives

Strategy and instruments
- Rise of general ideological preference for indirect and procedural instrumental strategies
- Tendency to incorporate multiple instruments in policy mix

Responsibilities and resources for implementation
- Rise of proportion of (relatively) independent and businesslike organised implementation organisations, including privatisation of water management tasks
- Tendency to rely on more than only judicial resources and to clarify responsibilities

While these fundamental sources of change agents are grouped by separate elements of the regime, we will use in the empirical research a more general categorisation of the more direct change agents that evolve from these fundamental ones:

a. European Union originated policy pressures;
b. National regime developments;
c. Problem pressures;
d. Various other pressures (e.g. rise of environmental NGO's).

These groups of change agents can be related to the development of regimes. The regime can be portrayed as moving from one stage to another. As long as one acknowledges that various intermediate and mixed situations are possible and probable, such stages can be a useful heuristic.

Simple regimes

The integration of regimes can be described in terms of extent and coherence. The extent of a regime is the scope of the uses and users that are regulated by one or more of the regime's elements. The coherence is about the degree to which these elements fit together. Simple regimes regulate only one resources use or user. It's the way -- in theory previously unregulated -- resources begin to be a subject of regimes. Furthermore, relatively singular (or simple) regimes (one level, one governing actor, one problem aspect -- e.g. a certain use or user -- one instrument, one implementing agency) will not be in need of coherence. Only after some growth in complexity does coherence become a relevant concept. But then, it is by no means a logical follow up. Complex but fragmented regimes are empirically quite common.[4] While more complexity is part of a stream of societal developments that seem ever to increase as time goes by, both coherence and fragmentation seem to be common developments.

Change toward complexity

When we speak of complexity it means that regimes can be characterised by multiple formats in most of their elements. A regime becomes more complex when more layers and scales are involved, more actors are involved, more perceptions of the problem and accompanying goals are involved, more instruments are part of the policy mix and more organisations share responsibilities for implementation. The most eminent feature is the gradual increase of the domain of the regime, that is the uses and users regulated by one or more parts of the regime. With it also comes an increase in relevant property and use rights. We will refer to this crucial variable as the regime's *extent*. Regimes with an insufficient extent are by definition weak as guardians of sustainable use, while some relevant parts of the domain go unregulated.

Figure 2.2*: Regime developments*

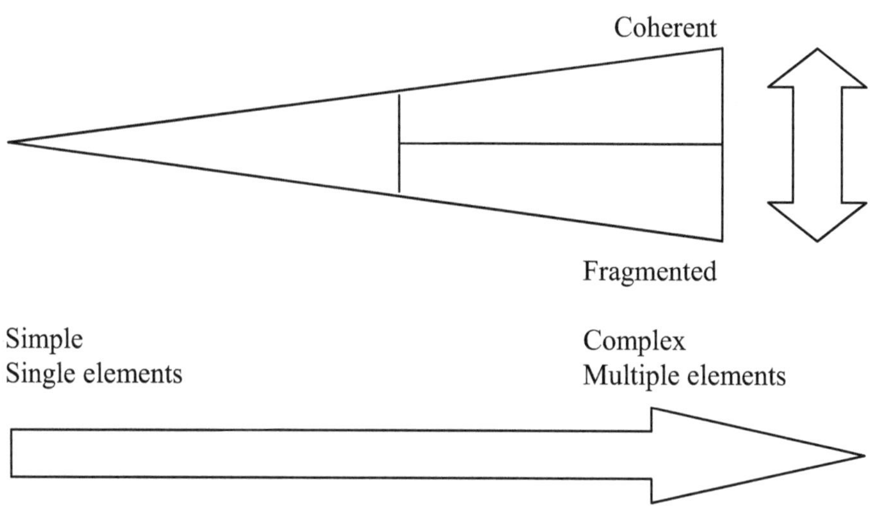

Complexity as such is thus not wrong. Most of the time, growing complexity is an answer to real needs and developments. As a matter of fact, societies

4 In fact, while integration has clear theoretical advantages, it comes at a price. Every form of integration creates the need for additional interaction and increases transaction costs, at least initially.

generally grew into more complexity during most of modern times. This sector is no exception to that general course of development. A growing complexity in governance can be viewed as a logical adaptation to that development. Many external change agents, such as technological developments, add new scales, new actors, new problem perceptions, new instruments, and new responsibilities to the existing ones. This leads to the first hypothesis:

1. Most change agents (in the period and context of our cases) will lead to more differentiation in the regime (resulting in more complex regimes).

Change toward integration

While the growth of complexity in water management regimes seems a fairly straightforward part of a more general development in society, integration as a development is not. (See Figure 2.2) While the term 'integration' is common in most policy papers (e.g. 'integrated water management'), in this project we choose to use the term coherence instead, for the reason that, in most policy papers the term integration (e.g. in 'integrated water management') is used in a sense that implicitly or explicitly includes an increase in the domain of the regime, the extent to all relevant users and uses. Therefore, we believe that integration as it is used in the policy sphere is a combination of what we call *extent and coherence*. For the sake of conceptual clarity and the possibility to adapt to the meaning of the term integration in policy practice, we use these terms further when appropriate, and reserve 'integration' for the combination of the two.

The central focus in our research is on the *changes* of the regime towards more integration (extent and coherence). This variable is the dependent variable in our most important research question and the independent variable in the other. Changes are placed at the forefront here, because in the case studies these transitions (towards more extent and coherence) themselves are the focus of attention. It was not always important to sketch all aspects of the regime, but only those that are necessary to understand what the observed changes mean in the context of the rest of the regime. Before we began, we expected that non-trivial changes (even if they involve changes towards 'consensual management' or the like) would often involve some kind of conflict, struggle or manipulation, with losers also involved. So, though it is not impossible that changes in problem perceptions (in combination with the existing 'regulative system' of property and use rights) invoke a real consensus that everybody is better off with more integration, this certainly need not be the case. It is even likely in such a case that there has been some previous struggle about the problem perception itself. Even though these changes are heading towards integration, the change process itself will often involve overt or hidden conflicts (as could be observed in several of the case studies). Such opposition can also lead to 'failed' or partial regime shifts towards more coherence, when changes in one element of the regime are encapsulated, rather than followed by matching changes in the other elements of the regime.

The resource regime consists of two components and their mutual relationships: the regulative system of property and use rights and the public governance system, with its five elements discussed in section 2.2.1. Consequently we discern three forms of coherence:

1. the internal coherence of the public governance component of the regime,
2. the internal coherence of the property rights component of the regime, and
3. the external coherence between the public governance and property rights components.

By *coherence of the public governance* component we mean the following. When more than one layer of government is dealing with the same water resource (as is often the case), then coherence means inter alia that the activities of these layers of government are recognised as mutually dependent and influencing each others' effects. When more than one actor or target group is involved in the policy, coherence means that there is a substantial degree of interaction in the policy network. When more than one use or user is causing the unsustainable problem, coherence means that the various resulting objectives are analysed in one framework so that deliberate choices can be made if and when goals are conflicting. The same holds for instrumental strategies that are used to attain the different objectives, as well as for the different instruments in a mix to attain one of these objectives. To conclude, coherence of the organisation of implementation means that responsibilities and resources of various persons or organisations that are to contribute to the application of the policy are co-ordinated, or these actors themselves are co-ordinated.

The *internal coherence of the property rights system* is threatened when property or use rights are given to actors for uses that decrease the possibility for uses that were already granted to others. This can have several background reasons. Sometimes use rights that were long seen as non-rival and thus compatible can become rival ones by a drastic increase in use, or by the use of new techniques. The internal coherence of the property rights system is thus the degree to which the interdependencies in the water system and its management that occur in reality are reflected within and between the property and use rights. The essence of this variable is that property and use rights of one actor do not inherently or under the given circumstances make it unavoidable to clash with other rights and/or with the sustainability of the resource, without external intervention to prevent it.

The two components of the resource regime lack *external coherence* in the first place in case of a wrong match between the actors targeted by the public governance system and the actors with relevant property and use rights. In the second place a mismatch of the goods and services involved in both systems might also lead to a lack of coherence and thus a possible form of change towards more coherence.

A change towards more coherence will occur only when relevant actors acknowledge that coherence is necessary to prevent further deterioration of the resource. That means that coherence is not a spontaneous development, but has a deliberate character. This means that coherence typically stems from special change agents that demand some form of coherence. A 'coherence' agent can, for instance, consist of the recognition of the interaction between multiple water uses or of a European Union directive that demands multilevel co-operation in water resource management planning. Also, international and inter-local learning is a possible change agent. Unlike an increase in complexity, then, developments in the direction of more coherence need some sort of deliberate attempt by motivated actors. This all leads to formulation of the next hypothesis:

2. Other external change agents of a specific nature (see above) can also lead to coherence in one or some elements of the regime, but only in combination with deliberate attempts of motivated actors (ultimately resulting in coherent regimes or in 'failed' regime shifts with encapsulated initial changes).

Conditions for coherence

Above, we stated that regime changes in the direction of more coherence are not spontaneous and typically will need deliberate attempts of motivated actors if they are to occur. This leaves open the question of the conditions under which such attempts or change agents will be relatively successful. Knoepfel (1995: 202-203), for instance, assumes that integration and co-operation between two or more agencies will work better when these agencies each have fewer separate tasks and goals and the agencies are matrix organisations. The second question of our research asks for the change agents and conditions for change. We elaborated that in three directions: the change agents (see hypotheses 1 and 2 above), the conditions for successful changes (hypothesis 3 below) and the more detailed form of the changes (hypothesis 4). Regime changes as such are often the result of processes that are not intended to produce a change in water management, or to do so, but in many cases for other reasons than water management purposes alone. The interactions of actors with their resources and within an institutional context produce the initial change. Behind these dynamics, more external change agents can often be recognised, which in turn affect the motives and resources of the actors involved. For the specific kind of changes that can be labelled 'moves towards more coherence', the expectation is that these changes will not occur unless there is some deliberate attempt by motivated actors towards coherence. The reason is that real coherence implies not only a reshuffling of power etcetera (that can also be produced at random by various external developments) but precisely that this reshuffling is NOT at random, but leads to the (in a world of randomised pressures) unlikely effect of more coherence. In our perspective, coherence will often meet resistance, like any other major change in relationships. So, besides 'losers', 'proponents' should probably also be identifiable (hypothesis 2). Otherwise changes will more often lead only to an increased extent and thus increased complexity (hypothesis 1).

The next idea that we wanted to test in our research is whether attempts to integrate need the same sort of favourable preconditions as some other forms of co-ordinated collective action. We stipulated five of these in hypothesis 3. In the comparative analysis in chapter 9 we shall try to discern patterns of conditions that accompany 'successes' in regime change towards more coherence. Change agents and conditions belong to the same set of causal factors. We distinguish them for the reason that the 'conditions' are often forgotten. Causal explanations are often sought in the form of 'new' and 'provocative' factors that are labelled as the 'causes'. In reality, this image of causality often forgets about the array of factors to which the analyst is used as being the 'normal' status (causal factors that one is inclined to forget about). It might then be delusory to think that these 'causes' really are the complete explanation of what happens. A simple example may clarify this. When a fire burns a house and one seeks the cause, one will be looking for sources of fire (e.g. an electrical short circuit) and exceptional forms of flammable material (e.g. a leaking cooking gas container). That there is a great deal of flammable material in a normal house and sufficient oxygen will be considered 'normal' or even not considered at all, while these factors are, of course, as essential as the previous ones.

In our cases, the division between the 'extraordinary' causes and the 'normal' conditions, that together form the 'causal set', are not as clear cut as in the example. Nevertheless, in the case studies in this book similar change agents sometimes also set in motion a development towards coherence and sometimes they don't. Compare similar seeds sown in different seedbeds. For us, the reasons why similar problem pressures all over Europe and similar EU and even national developments have dissimilar effects on regimes at the case level are interesting. Here the 'conditions' enter the picture. We hypothesise the following relationships:

3. Attempts to change regimes into a more coherent status will have relatively more success when:
– There is already a longer *tradition of co-operation* in the water management sector.
– There is a common understanding that the counteracting (side) effects of non-integrated water management harm sustainability and that this sooner or later will have to be stopped anyhow (*joint problem*).
– There is a notion of possible joint gains from coherence, so-called 'win-win situations' (*joint opportunities*).
– There is a credible threat of a dominant actor accumulating power and altering the public governance pattern in his interest when no solution is reached (*credible alternative threat*).
– There are well functioning institutions that provide fertile ground for coherence attempts (*institutional interfaces*).

The indicators for the relevant conditions used are:

3a Tradition of co-operation
- a dominant policy ideology that supports integration
- positive examples of integration known by the actors involved
- mutual respect and trust in 'fair play' of the actors involved

3b Joint problem
- knowledge bases in the form of reports and statements by respected sources on resource deterioration due to fragmentation
- information symmetry between the actors involved on these points
- a sense of responsibility for the future with the actors involved

3c Joint opportunities
- knowledge bases from respected sources on opportunities stemming from more integration
- information symmetry between the actors involved on these points
- a sense of respect for each others' interests among the actors involved

3d Credible alternative threat
- sufficient imbalance of power favouring a dominant actor (government?) to enable unilateral action
- information on alternative options to 'solve' the problem from the perspective of the dominant's actor's perspective
- alternative option has more severe consequences for the other stakeholders than the specific form of integration would have

3e Institutional interfaces
(not all indicators below are equally important to all forms of integration)
- clarity of assigned responsibilities (to prevent territorial battles)
- free and alert mass media to induce awareness of challenges to the system
- legal or practical possibilities to protect negotiated compromises from continuous litigation
- actors, independent or within the administration, with solely process objectives (brokers)
- a small number of stakeholders or a strong representative organisation for the major groups of stakeholders to enable authoritative a small number of interaction processes
- legal leeway for more integrative approaches
- official (not only laws, but also white papers and the like) policy guidelines to achieve more integration in water management

The devil is in the details

The following idea stems from our scepticism regarding the nature of many developments that are 'sold' as integrated management. We want everybody to be aware of the possibility that what on the surface looks like coherence may in fact have not much promise for sustainability in use, since the more detailed forms that the changes can take can actually be harmful. Here, the division between use-driven changes and protection-driven changes is extremely important. If new aspects of the problem gain recognition and are included in the weighting of objectives, it matters whether these are sustainability-related ones (often protection-driven) or not. The characteristics of the regime elements not initially affected can encapsulate the initial changes, even to the extent that these are effectively neutralised: "plus ça change, plus ça reste la même chose". This aspect was not so much a variable in the comparative analysis than a point of attention in the case studies.

Naturally, the dimensions of simplicity vs. complexity and coherence vs. fragmentation are characteristics of the regime that are of a rather general nature. Often, 'the devil is in the details'. When some actors find the network closed to them, it is important to learn who they are. And when new aspects of the problem gain recognition, it matters whether these are sustainability related ones or not. And so on. At a general level, no precise prediction can be established for these matters. However, our proposition of mutual adaptation of elements of governance leads to the expectation that the balances in the initially unaffected elements of governance are reflected in the way in which the greater or lesser 'seismic shocks' caused by the external change agents are absorbed in other elements. This leads to hypothesis 4, specified below:

4. The more detailed characteristics of regime changes to a large degree reflect the balances in other elements of the regime that were not directly influenced by the change agent(s) initially (including property rights).

Summary

Figure 2.3 shows the position of the four hypotheses. The first two hypotheses state that change agents will normally lead to more complexity and that it will need 'deliberate attempts of motivated actors' to turn change agents' influence into a change towards more coherence. The conclusion whether or not this is true has political relevance, for it would mean that even when circumstances seem to demand more coherence, one can by no means trust this to evolve as a sort of automatic adaptation. The background to this is that real coherence will tend to 'hurt' somewhere in the system. For the case studies, these hypotheses are relative easy to assess. Is it true or not that in the case (in fact, in each of the 'story lines') changes towards more coherence always involved some of these 'attempts by motivated actors'?

The third hypothesis specifies the 'favourable conditions' for 'regime changes towards more integration'. This hypothesis will lie at the heart of the comparative analysis of the cases. In this analysis we relate the degree and form of integration to the specified conditions as assessed by the researchers.

The fourth hypothesis demands that the case study should pay attention not only to the degree of integration, but also to the details of its elaboration. The expectation here is that these details can be understood on the basis of the characteristics of the regime elements that are not directly affected from the beginning. This hypothesis can only be assessed on the basis of the historical descriptions of the changes involved. The number of possible aspects (details) is simply too big to make a uniform list of indicators for all case studies. So this expectation is not dealt with in this comparative analysis, but forms the theoretical background of many of the observations made in the case studies.

Figure 2.3*: Hypotheses scheme*

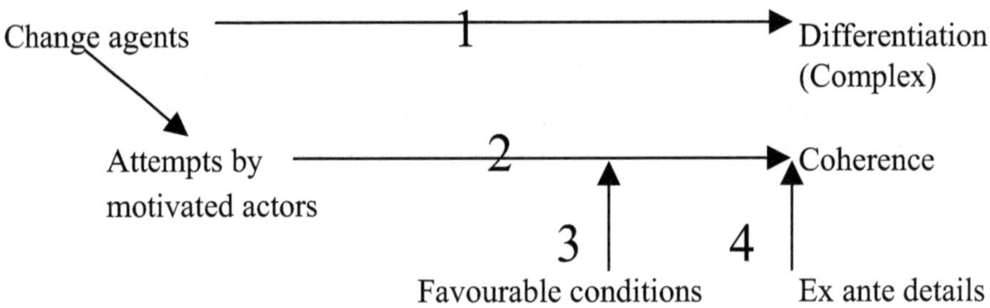

2.4 The sustainability implications of institutional resource regimes

The sustainability of a given institutional resource regime depends on its property and use rights component, the public governance component, and the interaction between these components. In the following discussion we therefore first delineate the sustainability implications of different property arrangements before turning to the sustainability implications of the public governance component, and finally connecting the two components. In this discussion we do not deal with the *'extent'* aspect again. We see a sufficient extent of the regime as a precondition for a benign effect on the sustainability of the use, since non-regulated uses and users will tend to disrupt the regime effects on sustainable use.

Though, for various reasons, it is hard to give an overall assessment of the sustainability of the regime, it is less difficult to assess whether the concrete regime changes lead in the direction of more or less sustainability. The expectation that more integrated regimes will ceteris paribus perform better for sustainability is part of European and many member states, political ideology on water management. A more theoretical reasoning on which one can base this expectation is given in this section.

There are many indicators that can genuinely be considered to represent aspects of sustainability (cf. the 'good status' as specified in the EU water directive -- see Chapter 1). It is not the purpose of our research in this book to assess the overall sustainability of the resource use. The approach to dealing with this variable (-group) is the following: We start with the rivalries that are at stake in the case story or stories. In the first instance, the assessment of the changes in the sustainability of the resource use is limited to the natural/environmental indicators that are directly at stake in these rivalries. In the second instance, the social and economical development consequences of the changes in these indicators and/or the measures taken for this purpose are also taken into account. In the last instance, a marginal check is also performed to see whether the observed changes have important side effects on other natural resource/environmental indicators and whether these in turn have indirect social and economic consequences.

2.4.1 Sustainability and the property rights component

Two 'variables' influence the sustainability implications of the de facto distribution of different bundles of property rights. These variables are the extent of collective action problems between the appropriators from the resource (which influence the probability that they will be able to jointly manage the resource in a sustainable way), and the economic benefit of an unsustainable management of the resource vis-à-vis the economic benefit of its sustainable management.[5]

The extent of collective action problems between a group of appropriators depends on numerous factors that are well documented in the literature on common pool resources and collective action in general (Ostrom, 1998; Sandler, 1992). Among the important factors are the

[5] This argument is based on the assumption that decision makers behave as homo oeconomicus. This is not always the case, of course. Individuals frequently behave in an environmentally conscious manner, even if it is not in their best economic interest. This willingness to behave altruistically, however, tends to diminish when it is associated with increasing costs (foregone benefit).

numbers of appropriators, the homogeneity of their interests, the presence of leadership, the technologies of appropriation, and, of course, the physical characteristics of the resource. The degree to which they are supported by the property rights arrangements, we call this the *'internal coherence of the property rights component'*.

The economic benefit of an unsustainable management of the resource vis-à-vis the economic benefit of its sustainable management influences the probability that appropriators will be interested in its sustainable management. The difference between the two economic outcomes is a function of market prices[6], of resource characteristics in terms of renewal rates, for instance, and of the discount rate applied by appropriators. In general, a substantially higher economic benefit of an unsustainable resource use vis-à-vis the economic benefit of a sustainable resource use is always likely to be associated with a less sustainable outcome.

The combination of these two variables has interesting implications for the sustainability of different distributions of property rights. If one considers a case in which the economic benefit of an unsustainable use of a resource is smaller than or relatively equal to the economic benefit of its sustainable use, a small extent of collective action problems among the appropriators will likely lead to a more sustainable outcome. An example of such a case is, for instance, the use of a fishing pond for fishing. Here, fewer appropriators, ceteris paribus, are more likely to be able to agree on sustainable harvesting rates than a large group of appropriators, which automatically faces higher transaction costs in terms of negotiating, monitoring, and enforcing rules.

If one considers a case in which the economic benefit of an unsustainable use of a resource is much larger than the economic benefit of its sustainable use, however, few and small collective action problems are likely to mean that the unsustainable use is most likely to be achieved. An illustration of such a case is, for instance, the existence of the fishing pond in Paris, where a small number of appropriators probably are going to be much more efficient in draining the pond and selling it to a developer. Thus, few and small collective action problems might actually lead to less sustainable outcomes of specific distributions of property rights if the unsustainable use of a resource is economically much more attractive than its sustainable use. Of course, this would be a typical case where some interference from the public governance component would be important. Nevertheless a certain degree of coherence of the property rights component can in many cases be an important asset for a more sustainable resource use.

6 One of the (indirectly) relevant market prices is the price of money, i.e. the interest rate, which determines the opportunity costs of foregone economic benefit.

2.4.2 Sustainability and the public governance component

On the public governance side of an institutional resource regime, its sustainability implications depend mainly on three variables: the validity of the 'policy theory', the degree and quality of implementation, and the degree of coherence. Since we concentrate on the latter in this research, we shall stipulate implications of the first two only briefly.

Validity. The optimal mix between the public policy and property rights aspects of the regime mix could be regarded as an aspect of the validity of the policy theory ('policy design'). The concept of 'policy theory' (Hoogerwerf, 1990) builds on the idea that policy can be regarded as an attempt to attain certain goals with certain means in a certain time perspective. A policy builds, explicitly or implicitly, on sets of assumptions regarding the causal relationships between variables in the targeted policy field, relationships between 'chosen and newly induced causes' and 'desired effects' (sometimes called final relationships) and relationships between general values and more specific standards and objectives. These assumptions can be more or less valid from the perspective of a certain policy objective, like sustainable resource use.

Furthermore, the aspects of the policy sector that the policy theory should reckon with include not only the causes, features and effects of the physical water resource use, but also the pattern of property rights that is connected to these uses. To design a policy that improves the use from a sustainability perspective, the policy has to reckon with what is already there as 'self-governing' capacity and what is lacking. An additional important aspect is the instrumental validity of the policy: Are the chosen strategies and instruments in principle, when implemented correctly, capable of achieving the desired results?

Implementation. The sets of incentives that are provided by the regime (including both property rights and public governance) are in many cases not self-executing. In order to be realised in practice, many need interaction processes in which they are applied. Often, these processes deal with the individualisation of general rules and subsidy or charge schemes or the enforcement of restrictions on resource use that are mandated by such rules or alternatively by the decisions of owners. When parts of the planned incentives are not implemented or implemented inadequately (meaning that the form in which implementation takes place decreases the incentive effect) even elements of the regime that in principle are complete and valid will prove to be insufficient to counteract unsustainable practices in resource use (cf. the 'contextual policy interaction theory', Bressers and Klok, 1988; Bressers and Kuks, 2001; Bressers, 2002).

Coherence. In the research on which this book is based, the Euwareness project, the most central characteristic for regimes to be successful in sustainable resource management has been coherence. This variable is thus discussed at some length here and is the central component of our expectations.[7] Coherence can apply to all elements of governance separately and mutually and similarly to the various property and use rights and the combination of the two components. So coherence can take various forms, all of which can contribute to the coherence of public governance as a whole, but each cannot necessarily provide decisive contributions to overall coherence on its own. To be able to judge what (combination of) aspects of coherence provides the stipulated condition for sustainable resource management, it is important to make clear what is to be expected from coherence. In our opinion, the principal benefit of coherence of the public governance component is to prevent negative side effects from some elements on the positive effects of other elements on the sustainability of resource management and, simultaneously, to obtain extra opportunities for productive co-operation of various elements of the regime.[8]

On the one hand, this means that 'coherence' cannot be conceived of as a single best way of dealing with these interactions. For that, the multiple possibilities are simply too overwhelming. Various forms of coherence might provide various partial solutions to existing problems. On the other hand, it is clear that there are even more ways in which a lack of coherence can be detrimental to the regime's effect on sustainable use. All in all, we predict that substantial steps toward more coherence will decrease specific forms of unsustainable use and that even in cases of a valid policy design and good implementation, certain specified forms of a lack of coherence will cause flaws in the sustainability of resource use.

7 'Sufficient coherence' can also be conceived of as an aspect of the validity of the policy theory for the design features concerned and as an aspect of implementation for the process features concerned, but by doing so its emphasis would be lost.

8 As Ligteringen (1996, 1998) has shown, side-effects of policies can actually have a greater impact on the attainment of certain environmental goals than the specific policies designed to influence that aspect of the problem. She states that part of these side effects occur indirectly through influencing major societal developments that themselves have a substantial effect on sustainability. Because of these indirect influences there is a multitude of possible effects of the various elements of the regime on the sustainability of the resource use. It is also possible that the effects of one instrument are counteracted directly or indirectly by the effects of other regime characteristics.

To determine any additional impact of better coherence, we need to go back to the implementation condition. In a dynamic context and surrounded by uncertainties regarding the problem and possible solutions, an adaptive and learning form of policy implementation is important. Such a policy style depends on the uneasy combination of both pluralism (e.g. allowing complexity, openness, differentiation) to give room for challenging stimuli, and co-ordination (e.g. coherence, consensus seeking devices), to be able to produce new solutions on the basis of these challenges (Arentsen, Bressers and O'Toole, 2000). Cases in which consensus seeking devices are strong, but pluralism is restricted to 'insiders', like the 'iron triangle' that agriculture has been for a long time in many countries, are, however, vulnerable to becoming too closed for external stimuli that would provide the system with timely incentives to adapt and learn. Cases in which pluralism is strong, but no strong devices press for a co-ordinated agreement (like some examples of US environmental policy) lack the incentives to explore win-win situations or at least profitable trade-offs.

For our analysis, this means that the role of coherence is connected to the ability to incorporate complexity in productive collective action. We assume that more coherence correlates with less unproductive constellations of goals, information and power of the actors involved in the implementation process. With more coherence in the public governance component of the regime, the goals of the implementers and target groups involved in the implementation process can be expected to be less likely in discord. All elements of a more coherent regime can be assumed to contribute to a lowered degree of experienced uncertainty, an increase in information exchange, and a lower degree of distrust. Coherence also means that there will typically be fewer possibilities for target groups to play implementers off against each other and more standard operation procedures for the solution of conflict.

This implies that a more coherent public governance component of the regime can outperform a regime with an equal degree of extent, but more fragmentation. This is expected to be the case, not only through the direct effects of more mutually reinforcing and less mutually destructive side effects on the resource use, but also through indirect effects on the quality of the implementation process.

2.4.3 Sustainability and the combination of both components

Given this understanding of the implications of the property rights and the public governance components for the sustainability of institutional resource regimes, we now turn to the implications of public governance intervention for property rights. We base this discussion on the notion that (almost any) policy translates into an intervention and change in de facto private property rights. The sustainability implications of an institutional resource regime at any point in time, therefore, are to a large extent determined by how policies structure, i.e. create and influence, property arrangements. To what extent is such public governance intervention in property arrangements necessary or desirable for the sustainability of an institutional resource regime? The answer to this question depends first, of course, on the potential for sustainability of the given property arrangement, as discussed in 2.4.1. In addition, the answer depends on the potential of governance intervention to improve the sustainability of the given resource management. This potential of public governance intervention to lead to an improvement, depends in turn on the sustainability potential of the governance pattern discussed above. Thus, the potential of public governance intervention in property rights to improve the sustainability of resource use depends on the validity of the policy theory, the probability of (adequate) implementation of the policy, and sufficient coherence of a given policy. In general, we can see that the environmentally desirable degree of public governance intervention is higher, the higher the governance strategy scores on the three factors.

Combining the property rights side with the public governance side of an institutional resource regime, then, leads to the following statements:

1. The greater the validity of the policy theory, the higher the probability of correct implementation, and the better the coherence, the higher the environmentally desirable degree of public governance intervention.
2. The greater the economic benefit of an unsustainable resource use vis-à-vis the economic benefit of a sustainable one, or the more there are large and numerous collective action problems among property and use rights holders, the greater the environmentally desirable degree of public governance intervention.

Please note that although these statements are formulated in terms of greater desirability of public governance intervention, the opposite cases imply lesser desirability of public governance intervention and the expectations could just as well be formulated in that way, too. In other words, this analysis does not imply that public governance intervention will always have a positive effect on the sustainability of an institutional resource regime and therefore is always desirable and necessary. Indeed, incidents combining a low validity of the policy theory, incomplete and incorrect implementation, and absence of coherence are the clearest cases where the opposite is true. When we speak of the external coherence between the public governance and the property rights components of the regime we point not only to the appropriate degree of public governance intervention. Instead, we mean also and foremost the degree to which the right connections are made between the elements of public governance (e.g. policy instruments) and the relevant aspects of the property rights arrangements (e.g. whether the right property right owners are targeted by the policy instruments).

2.5 Overview of hypotheses

After having dealt with all hypotheses we can now give an overview. There are three groups of variables, linked by the central relationships in the research questions.

A. Change agents and conditions →
B. Regime changes →
C. Effects on sustainability of resource use

Our hypotheses and expectations are as follows:

Hypotheses on regime change towards integration

1. Most change agents (in the period and context of our cases) will lead to more differentiation in the regime (resulting in more complex regimes with a greater degree of *extent*).
2. Other external change agents of a specific nature (see above) can also lead to *coherence* in or between one or some elements of the regime, but only in combination with deliberate attempts of motivated actors (ultimately resulting in coherent regimes or in 'failed' regime shifts with encapsulated initial changes).
3. Attempts to change regimes into a more integrated status will have relatively more success when:
   - There is already a longer *tradition of co-operation* in the water management sector.
   - There is a common understanding that the counteracting (side) effects of non-integrated water management harm sustainability and that this sooner or later will have to be stopped anyhow (*joint problem*).
   - There is a notion of possible joint gains from coherence, so-called 'win-win-situations' (*joint opportunities*).
   - There is a credible threat of a dominant actor accumulating power and altering the public governance pattern in his interest if no solution is reached (*credible alternative threat*).
   - There are well functioning institutions that provide fertile ground for coherence attempts (*institutional interfaces*).

4. The more *detailed characteristics* of regime changes to a large extent reflect the balances in other elements of the regime that initially were not directly influenced by the change agent(s) (including property rights).

Expectations on the sustainability of institutional resource regimes

5. Regimes with a deficient *extent* will be more likely to lead to degradation of water resources or inability to protect the ecological functions of the water resource, than regimes with a larger extent.
6. Regimes with a large 'extent', but with low *coherence* will be more likely to lead to degradation of water resources or inability to protect the ecological functions of the water resource, than regimes with a similar extent but a higher degree of coherence.

2.6 Case study design

The case studies and case study comparison is not the whole Euwareness study. Besides to the case studies 'country screenings' are also done, describing and assessing policy and property right developments on the national level over a prolonged period (Kissling-Näf and Kuks, 2004). The methodological remarks below, however, are attuned to the case studies and the way the comparison is made in this report. The case selection will be dealt with first, then the design of the case studies. The chapter concludes with some remarks about the way the case study researchers assessed the variables.

2.6.1 Selection of cases

In the research on which this book is based, the Euwareness project, two cases are studied in each of the six participating countries. In a first period of six months a first set of six cases (one in each country) was studied. The second set of six cases followed in a subsequent period of six months. The main criteria for the selection of the cases were:

- The demarcation of a case should follow the hydrological and geographical boundaries of a water basin at a regional scale or with a tributary character.
- We have been looking for cases of rivalry between heterogeneous / homogeneous uses/users of the same water resource. We preferred cases where several rivalries show up to allow intra-case analysis. It was not necessary that these rivalries are manifest in the whole case area, they might also be at stake in just a part of the case area.
- There was a preference for cases where not only public ownership but also private ownership of water resources could be found.

- Cases should be selected on the presence of at least attempts to obtain transitions towards more coherence during the last two decades.

The sample of case studies is based on a combination of similarities and differences. In some respects, it seeks similarities (e.g. medium size river basins) that define boundaries of the research subject. In some respects, it deliberately encompasses different situations (e.g. 'wet' cases and 'dry' cases). But the most significant decision has to do with how the cases relate to the main three variable-groups, since these relations influence the inferences that can be drawn about the hypotheses that relate to these variables (see 2.5).

There are various modes of sample selection, depending on the sort of inferences one wants to make. On the surface, the last mentioned criterion, namely that there should be the ex ante impression that a serious attempt to attain more integration in the regime took place in the proposed case, looks similar to the other (similarity) criteria. Nevertheless, this criterion is a combination of an extreme case sampling strategy and a random variation driven strategy (cf. Patton, 1980). It is extreme case sampling in the sense that it leaves out all possible cases where there is no ex ante evidence that attempts towards more integration have been made. The implication of this is that if we don't find improved factual ('real') integration in our cases, the chances are slim that we shall find it on any large scale outside of our sample.

It is also a random variation strategy though, since any attempt to attain more integration surely doesn't imply its success in close observation. On the contrary, we expect to see anything, from major improvements to only symbolic alterations and everything in between, due to the various conditions of the case. To re-use a metaphor from above: we confine ourselves to cases where seeds have been sown, in the expectation that these will be shown to bear fruits to very divergent degrees. This gives us the opportunity to make an inventory of change agents observed and test expectations about beneficial conditions. On a separate case level, the disadvantage is that in case that in practice little or no regime change towards more integration could be shown, it is not possible to look for sustainability effects of these non-existent regime changes. Nevertheless, on a comparative level we'll find some variation in the independent variable, with the hypothesis to be tested that improved integration will show connections with some improved aspects of sustainable resource use.

2.6.2 Case study protocol

The case studies had two stages. The first one is a descriptive one, in which the emphasis lies on the story or stories to be told. The second one is an analytical one in which the values of the variables are assessed that play a role in the theory that is used in the intra- and/or inter-case comparisons to arrive at an answer to the research questions of the project (Dente, Fareri and Ligteringen, 1998). In many cases, the case study will contain not only one story of regime change, but more than one. This may imply developments that can be seen as partial coherence in geographical sub-units of the case study territory or between certain aspects of the resource use but not between others. Our proposal was not to submerge these sub-stories and force them into one over-all case description, but to pay separate attention to them against the background of descriptions of the more general case situation and development. So some of the cases contain more than one more or less independent development or 'story'. In these cases, *sub-cases* may be discerned. There is only one case-story under the following conditions:

- If there is only one (major) or at least only one selected rivalry;
- If there is only one line of development or only one aspect with which the regime has changed;
- If the regime changes observed are highly interdependent; and/or
- If the rivalries in the case are highly interdependent.

If none of these statements hold true, we discerned separate subcases when analysing the variables and hypotheses in chapter 4. A subcase is then a set of observations for which the above criteria *do* hold.[9] In many instances, this also meant that not only regimes on the water resource, but also regimes on land use, nature protection and other natural resources (e.g. fish) were at stake.

The descriptive part of the case study follows mostly a historical logic. If there is more than one important story line or sub-case, these are dealt with separately. From the beginning, though, the stories are selected ones. Only those developments that are relevant in the light of the variables and theories of the project are worth elaboration. This makes it easier for the second part of the case study to be linked with the descriptive part. Since all teams had extensive experience with case studies before, the case study protocol used did not further elaborate on the descriptive part.

[9] Compare a detective story in which more than one murder takes place. If these are interconnected it makes no sense, when analysing the plot, to divide them into subcases, but if they are just connected by the fact that they take place in more or less the same period, they will probably have quite different plots that require separate attention when analysed.

The analytical part consists of the assessment of relevant variables (translating 'real life' observations into theoretical language) and the inferences and conclusions that can be based on these variables and their relationships. For the case study protocol, the identification of the key variables to be assessed and their indicators was most important. Since subcases are treated as equal cases in the analysis of the assessments of the relevant variables, cases that are split into subcases are in a sense over-represented in the data for the comparative analysis that is presented in Chapter 9. Therefore, we also constructed a 'weighted database' in which all cases were assigned four units of research. That means that when a case is not split into 2 or 4 subcases but analysed as a single case, that case was included fourfold in the 'weighted database'. All the analyses presented in chapter 9 were also done with this 'weighted database'. Commonly, though, the results did not differ.

While it was seldom possible to give a complete overall assessment of the status of the regime in our case studies (too many aspects would require attention), it was possible to assess the changes that have occurred in the research period of the case study. Do they simply add uses or users to the domain of the regime (extent), or are they also 'repairing' one or more mismatches (coherence), and if so which ones? Then, the choice of 'issues' or 'rivalries' determines what changes are relevant in the context of the case story/ies.

In our case studies, it has turned out that there is no escape from more or less artificial boundaries of the case as studied, boundaries that are set by the researchers. These boundaries exist not only in terms of geographical area or time period, but also may exclude certain issues.[10] This happened, for instance, with the rivalry between kayakists and fishers in both French cases. By approaching the object of study from the perspective of the rivalries it is possible to assess the impact of the changes on the regime's status in as far as it is relevant for the (sub-)case (more or less extent, more or less of various kinds of coherence).

2.6.3 From case studies to comparative analysis

As an aid to the comparative analysis, questionnaire forms were used for the case study researchers to fill in. These 'case study fact and assessment sheets' represent the variables and indicators of the theoretical model. Their purpose was to summarise the information in a uniform format so that the case information is comparable along the lines of the theoretical variables

[10] An example of a change that is not relevant to the case story is when national law sets general minimum quality standards when studying a clear mountain river where quality has never been any problem.

and hypotheses. The exercises of filling in the forms also proved very helpful in getting a grip on the case analysis itself.

Apart from the few short statements per variable ('key facts'), the researchers were asked to use a five-point scale to score the variables in order to make the cases comparable. Of course such a score is not a fact, but a judgement, much like marks are with school test papers. Therefore, we also wanted to know the most relevant facts they observed that they had in mind while scoring ('key facts'). While it might give a case study researcher an uncomfortable feeling to transform observations into scores, in fact, it gave them an influence on the way the case study comparison is made. For when comparing cases one always makes, explicitly or implicitly, these kinds of judgements on the rating of variables. We choose to do so explicitly.

The great advantage of this procedure is that the people who doe the assessments have extensive and intensive knowledge about the cases at hand, often even more than they described in the reports. In this way, we tried to combine the better of two worlds: the depth of information realised in extensive case studies and the clarity and overview of a data-matrix enabling all kinds of comparative analysis (cf. Patton, 1980). Compared with the direct, qualitative comparison of the case studies as reported, the approach diminishes the risk of bias that the comparative analyst is mislead by surprising, but anecdotal evidence of only one or two cases that is not representative of the relationships in the whole sample of cases.

A possible disadvantage is that different researchers could each interpret the variables a little differently. We tried to counter this possible disadvantage by providing the researchers with two completed 'pilot' 'case study fact and assessment sheets' as examples, namely on the two Italian case studies. We also presented an additional explanatory paper. After the receipt of the filled-in electronic forms from the case study researchers, we checked whether there was any reason to doubt correct interpretation (always partly on the basis of the key facts presented by the case researchers themselves). These -- occasional -- hesitations were then communicated to the case researchers. Sometimes, this resulted in changes in the assessments. Mostly, a further clarification and underpinning by additional 'key facts' could be given as an adequate response.

In addition, one might question whether the case study researchers were not unconsciously inclined to 'fix' the case by assessing the variables not really independently of each other but having the scores on dependent variables influenced by their assessment of independent variables or vice versa. The demanded association with mentioned 'key facts' already gave some protection against this. Furthermore, luckily, we were able to test this possible form of bias. In the theory, both the forces of the change agents and the conditions for change explain regime change. The latter are the less 'visible' elements of the causal set. In the case study reports, far more attention was paid to the various change agents than to the conditions. This is often even a large part of the story in the reports. If the suspected form of bias were real, then one could expect the variables of 'degree of regime change' and 'force of change agents' to be scored by the researchers in such a way that they would correlate strongly. But the opposite is true: the force of change agents proved afterwards to be far less strongly correlated with regime changes than the conditions are (see Chapter 9). This attests that the researchers assessed the variables independently on their own merits.

REFERENCES

Arentsen, Maarten J. and Rolf W. Künneke (1996) Economic organization and liberalization of the electricity industry, in: *Energy Policy*, Vol. 24, no. 6, pp. 541-552.

Arentsen, Maarten; Hans Bressers and Laurence O'Toole (2000) Institutional and policy responses to uncertainty in environmental policy: A comparison of Dutch and US styles, in: *Policy Studies Journal*, Vol. 28, No. 2, pp. 597-611.

Björk, Peder and Hans Johansson (2000) *Towards a governance theory: A state-centric approach*, paper IPSA Quebec, August 2000.

Barzel, Yoram (1989) *Economic analysis of property rights,* Cambridge: Cambridge University Press.

Baumgartner, Frank R. and Bryan D. Jones (1993) *Agendas and instability in American politics*, Chicago: University of Chicago Press.

Blomquist, William and Edella Schlager (1999) *Watershed management from the ground up*, paper APSA Atlanta.

Bressers, Hans and Pieter Jan Klok (1988) Fundamentals for a theory of policy instruments, in: *International Journal of Social Economics*, Vol. 15, No. 3/4, pp. 22-41.

Bressers, Hans and Laurence O'Toole (1995) Networks and water policy, Conclusions and implications for research, in: Hans Bressers, Laurence O'Toole and Jeremy Richardson, *Networks for water policy: A comparative perspective*, London: Frank Cass, pp. 197-217.

Bressers, Hans (2002) *Understanding the implementation of instruments; How to know what works, where, when and how*, paper international SUSGOV working group meeting, October 2002, Rome, Italy.

Bressers, Hans and Stefan Kuks (2001) Governance patronen als verbreding van het beleidsbegrip, in: *Beleidswetenschap*, Vol. 15, No. 1, pp. 76-103; also to be published as: What does governance mean? From concept to elaboration, in: Hans Th. A. Bressers and Walter A. Rosenbaum (Eds.) *Achieving sustainable development: The challenge of governance across social scales*, New York-Westpoint-London: Praeger (2004).

Bromley, Daniel W. (1989) Property Relations and Economic Development: The Other Land Reform, in: *World Development* Vol. 17, pp. 867-877.

Bromley, Daniel W. and Ian Hodge (1990) Private property rights and presumptive policy entitlements: Reconsidering the primises of rural policy, in: *European Review of agricultural economics,* Vol. 17, No. 1, pp. 197-214.

Bromley, Daniel W. (1991) *Environment and Economy. Property Rights and Public Policy*, Oxford UK /Cambridge USA: Blackwell.

Burns, Tom R. and Thomas Dietz (1996) Cultural Evolution: Social Rule Svstems, Selection and Human Agency, in: *International Sociology,* Vol. 7, pp. 259-283.

Dente, Bruno; Paolo Fareri and Josee Ligteringen (1998) A theoretical framework for case study analysis, in: Dente, Fareri and Ligteringen (eds.) *The waste and the backyard*, Dordrecht: Kluwer, pp. 197-223.

Devlin, Rose Anne and Quentin R. Grafton (1998) *Economic Rights and Environmental Wrongs. Property Rights for the Common Good,* Cheltenham: Elgar.

Dryzek, John S. (1987) *Rational ecology: Environment and political economy*, Oxford: Basil Blackwell.

Dryzek, John S. (1997) *The politics of the Earth: Environmental discourses*, New York: Oxford University Press.

Ensminger, Jean and Andrew Rutten (1991) The Political Economy of Changing Property Rights: Dismantling a Pastoral Commons, in: *American Ethnologist* Vol. 18 No. 4, pp. 41-57.

Feeny, David (1988) The Development of Property Rights in Land: A Comparative Study, in: Robert H. Bates, *Toward a Political Economy of Development. A Rational Choice Perspective,* Berkeley, Los Angeles and London: University of California Press.

Feeny, David; Berkes, Fikret and Bonnie McCay (1990) *The tragedy of the commons. Twenty-Two years later, in: Human ecology, Vol. 1, pp. 1 - 19.*

Fisher, Frank (1995) *Evaluating public policy*, Chicago: Nelson-Hall.

Fisher, Frank and John Forrester (eds.) (1993) *The argumentative turn in policy analysis and planning*, Durham: Duke University Press.

Furubotn, Eirik G. and Svetozar Pejovich (1975) Property Rights and Economic Theory: A Survey of Recent Literature, in: Henry G. Manne (ed.) *The Economics of Legal Relationships. Reading in the Theory of Property Rights,* St. Paul, New York, a/o.: West Publishing Company. pp. 53-65.

Hoogerwerf, A. (1990) Reconstructing policy theory, in: *Evaluation and program planning*, Vol. 13, No. 3, pp. 285-291.

Kasper, Wolfgang and Manfred Streit (1998) *Institutional Economics. Social Order and Public Policy*. Cheltenham: Edgar.

Kingdon, John (1995, 1st ed. 1984) *Agendas, alternatives and public policies*, New York: Harper Collins.

Kissling-Näf, Ingrid; Kuks, Stefan M.M. (eds.)(2004) *The Evolution of National Water Regimes in Europe. Transitions in water rights and water policies towards sustainability*, Dordrecht-Boston-London: Kluwer Academic Publishers.

Kooiman, Jan (1993) Findings, speculations and recommendations, in: Jan Kooiman (ed.) *Modern governance: New government – society interactions*, London etc.: Sage, pp. 249-262.

Knoepfel, Peter (1995) New institutional arrangements for a new generation of environmental policy instruments: Intra- and interpolicy cooperation, in: Bruno Dente (ed.) *Environmental policy in search of new instruments*, Dordrecht: Kluwer, pp. 197-233.

Libecap, Gary D. (1993) *Contracting for Property Rights,* Cambridge: Cambridge University Press.

Ligteringen, Josee (1996) The effects of public policies on household metabolism, in: Klaas Jan Noorman and Ton Schoot Uiterkamp, *Green households? Domestic consumers, environment and sustainability*, London: Earthscan, pp. 212-235.

Ligteringen, Josee (1998) *The feasibility of Dutch environmental policy instruments*, Enschede: Twente University Press.

Lynn Jr, Laurence E.; Carolyn J. Heinrich and Carolyn J. Hill (2000a) *The empirical study of governance: Theories, models and methods*, Washington DC: Georgetown University Press.

Lynn Jr., Laurence E.; Carolyn J. Herich and Carolyn J. Hill (2000b) Studying governance and public management: Challenges and prospects, in: *Journal of Public Administration and Theory*, Vol. 10, April 2000, pp. 233-261.

March, James G. and Johan P. Olson (1995) *Democratic governance*, New York, London: The free press.

Milward, H. Brinton and Keith G. Provan (1999) *How networks are governed* (unpublished paper).

North, Douglass (1989) Institutions and Economic Growth: A Historical Introduction, in: *World Development*, Vol. 17, No. 9, pp. 1319-1332.

Ostrom, Elinor (1990) *Governing the Commons, The evolution of institutions for collective action,* Cambridge: University Press.

Ostrom, Elinor (1992a) *Crafting Institutions for Self-Governing Irrigation Systems*, San Francisco, California: ICS Press Institute for Contemporary Studies.

Ostrom, Elinor (1992b) *The Rudiments of a Theory of the Origins, Survival, and Performance of Common-Property Institutions,* Workshop in Political Theory and Policy Analysis, Bloomington, Indiana.

Ostrom, Elinor (1994) *Neither Market nor State. Governance of Common-Pool Resources in the Twenty-First Century,* Paper presented at the IFPRI Lecture Series, Washington, D.C.

Ostrom, Elinor (1997) Private and Common Property, *The New Plagrave Dictionary of Law & Economics, Vol.* 3, pp. 424-432.

Ostrom, Elinor (1998) A Behavioral approach to the rational choice theory of collective action, in: *American Political Science Review*, Vol. 92, No. 1, pp. 1-22.

Ostrom, Elinor, Roy Gardner and James Walker (1994) *Rules, Games, & Common-pool Ressources,* Michigan: The University of Michigan Press.

O'Toole Jr., Laurence J. (2000) Research on policy implementation: Assessment and prospects, in: *Journal of Public Administration and Theory*, Vol. 10, April 2000, pp. 263-288.

Patton, Michael Quinn (1980) *Qualitative evaluation methods*, Beverly Hills / London: Sage.

Peters, B. Guy and John Pierre (1998) Governance without government? Rethinking public administration, in: *Journal of Public Administration and Theory*, Vol. 18, April 1998, pp. 223-243.

Pejovich, Svetozar (1975) Towards an Economic Theory of the Creation and Specification of Property Rights, in: Henry G. Manne (ed.) *The Economics of Legal Relationships. Readings in the Theory of Property Rights*, St. Paul, New York, a.o.: West Publishing Company, pp. 37-52.

Rhodes, R.A.W. (1997) Understanding Governance: Policy Networks, Governance, Reflexivity and Accountability, Open University Press.

Rhodes, R.A.W. (2000) Governance and public administration, in: Jon Pierre (ed.) *Debating governance*, Oxford: Oxford university press.

Rose, Richard (ed.) (1980) *Challenge to governance: Studies in overloaded polities*, Beverly Hills / London: Sage.

Rosenau, James N. (2000) *The governance of fragmegration: Neither a world republic nor a global interstate system*, paper IPSA Quebec, August 2000.

Sabatier, Paul A. and Hank C. Jenkins-Smith (1993) *Policy change and learning: An advocacy coalition approach*, Boulder: Westview.

Sabatier, Paul A. and Hank C. Jenkins-Smith (1999) The advocacy coalition framework: An assessment, in: Paul A. Sabatier (Ed.), *Theories of the policy process*, Boulder: Westview, pp. 117-168.

Sandler, Todd (1992) *Collective action. Theory and applications*, Ann Arbor: The University of Michigan Press.

Scharpf, Fritz W. (1997) *Games real actors play,* Boulder: Westview Press.

Schlager, Edella and Elinor Ostrom (1992) Property Rights Regimes, in: *Land Economics,* Vol. 68, pp. 250 ff.

Schön, Donald A. and Martin Rein (1994) *Frame reflection: Toward the resolution of intractable policy controversies,* New York: Basic Books.

Schwarz, Michiel and Michael Thompson (1990) *Divided we stand: redefining politics, technology and social choice*, New York: Harvester Wheatsheaf.

Smith, Andy (winter 1997) Studying multi-level governance, in: *Public Administration*, Vol. 75, pp. 711-729.

Stubblebine, Wm. Craig (1975) On Property Rights and Institutions, in: Henry G. Manne (ed.) *The Economics of Legal Relationships. Readings in the Theory of Property Rights*, St. Paul, New York, a.o.: West Publishing Company, pp. 11-22.

Thompson, Michael; Richard Ellis and Aaron Wildavsky (1990) *Cultural theory*, Boulder: Westview Press.

Young, Oran (1994) *International Governance. Protecting the Environment in a Stateless Society*, Itaca: Cornell University Press.

Zahariadis, Nikolaos (1999) Ambiguity, time, and multiple streams, in: Paul A. Sabatier (ed.) *Theories of the policy process*, Boulder: Westview, pp. 73-93.

Chapter 3

Harboring Water in a Crowded European Delta

The IJsselmeer and the Regge in the Netherlands

Dave Huitema* and Stefan Kuks**
**Vrije Universiteit Amsterdam (Netherlands)*
***University of Twente (Enschede-Netherlands)*

3.1 Introduction

The Netherlands is a relatively small, crowded country, located in the delta of three European river basins: the Rhine, the Meuse and the Scheldt. The country depends very much on transboundary inflows. Not only in terms of water quality and its vulnerability to upstream pollution sources, but also considering the country's dependence on over 75 percent of its total water resources coming from rivers abroad. Over 30 percent of the total surface area of the Netherlands lies below sea level, protected from the sea in the west and north by barriers of dunes and dykes. As much as 50 percent of the country's area is vulnerable to flooding from the sea or rivers. The Netherlands has a population of about 16 million inhabitants, an average of 470 inhabitants per km^2, giving it one of the highest population densities in the world. The highest concentrations are in the low-lying urban areas in the west of the country, which is the urbanized area including cities like Amsterdam, Rotterdam, The Hague and Utrecht. All Dutch urban areas together cover about 14 percent of the country's total surface. More than 50 percent of the country's area is agricultural land, and about 17 percent is water. Through Dutch history, there always has been a tension between urban, economic and agricultural development on the one hand, and the space naturally claimed by water in a delta area on the other. The need to protect the land from high water from rivers and sea, and the tradition of artificially draining low-lying areas, have given the country a complex

Hans Bressers and Stefan Kuks (eds.), Integrated governance and water basin management. Conditions for regime change towards sustainability, 59-98.

hydraulic infrastructure. Through the ages the flow and level of almost every water body in the country have been subject to human control.

Although the Netherlands is internationally appreciated for its great expertise in water engineering solutions, the country also struggled with critics from various groups in society on the predominant civil engineering orientation in Dutch water management. In fact these criticisms, which started to find expression in the 1960s and 1970s, politicized water management and initiated a debate in society on water values. The critics placed a greater value on the meaning of 'open' water for recreation, nature conservation, water storage, and the experience of unspoiled space in an already crowded and highly planned country. The intensification of agriculture since the 1950s and the related canalization of natural water courses and deterioration of landscape were also increasingly criticized. These criticisms resulted in a gradual greening of water engineering. In the 1980s, nature conservation and ecosystem protection became focal points of water management. The focus changed not only at the national level, where the state water authority [Rijkswaterstaat] takes responsibility for the main water bodies throughout the country, but also at the regional level, where for many centuries water boards [waterschappen] have been responsible for regional water management. The predominant focus of water boards on drainage for agricultural and urban development turned into an integral water system approach, combining quality and quantity aspects of water management, acknowledging the interrelatedness of surface and groundwater, and taking into account the value of water for the surrounding ecosystem (Grijns and Wisserhof, 1992; Snijdelaar, 1993; Disco, 1998).

During the 1990s the Dutch scope of water management widened even further. River floods in 1992 and 1995 and high water in 1998, causing large scale evacuations of inhabitants and enormous damage to property, triggered the awareness that the Dutch hydraulic infrastructure is facing increasing problems in keeping high water under control. Although climate change causes a rising sea level and higher rainfall peaks, which strengthens the dynamics of river basins, it is especially human interventions in river basins that have caused an enormous loss of space which is needed to store excessive quantities of water. Land use decisions of the past have taken insufficient into account that a delta area cannot completely rely on artificial control of water levels, and therefore needs space along rivers to allow river levels rise and fall in a more natural way. Water management has always served land use decision making, but now it should be the other way around: land use planning should consider water as a guiding principle for decision making. Nowadays, the challenge faced by the Netherlands is to harbor water in a crowded delta area (Commissie WB21).

In section 2 of this chapter we consider the most important transitions towards integration in national water management during the past decades. In the following sections we question what has been the practice of integrated water management in two water basins as cases. Section 3 is about the IJsselmeer area in the heart of the Netherlands, an area where an interesting variety of water management dilemmas is revealed. Water management in this area is directed by the state water authority, although decentral water authorities are becoming increasingly influential. Section 4 is about the Regge river basin in the east of the Netherlands, a case which is considered to be one of the earliest examples of integrated water management at the regional level. Water management in this river basin is dominated by a regional water authority (water board), called 'Waterschap Regge en Dinkel'. In section 5 we discuss the conditions under which water basin regimes could become more integrated. What could be the influence of national and European conditions in achieving regime transitions at water basin scale? We also discuss the difficulties that integration attempts are facing and how easily integration might turn into fragmentation. The chapter ends with a concluding section.

3.2 Transitions towards integration in the national water regime

In this book we are interested in transformation processes of water basin regimes, especially how they transform from complex into integrated regimes. In the case of the Netherlands such a transformation process started in the 1960s, with important transitions in the national water regime around 1968/69, 1985 and 1995. We shall consider these three transitions more closely to determine what kind of integration has been achieved. We use the indicators for integration as they are defined in this book, which means we look at the increase of extent and coherence. Although the identification of transitional moments is based on the appearance of important integration attempts at the national level in that year, we are aware that such attempts are part of a longer transformation process which generally started several years before and also continues for a few years after the identified transitional moment. For each transition, therefore, we describe the multi-year transformation process in which it is embedded.

3.2.1 The 1968 and 1969 transition

Until the 1950s the Dutch national water regime had little complexity. Water management was mainly a matter of flood prevention and water level control. During the 1950s and 1960s the complexity increased, and at the end of the 1960s first attempts at integrated water management could be perceived. These attempts included the adoption of a first national water policy plan in 1968 and the adoption of a Surface Water Pollution Act in 1969. The water quantity oriented policy plan not only focused on flood protection and drainage (water security), but also on water scarcity and the rival demands of water supply, agriculture and navigation. The Surface Water Pollution Act involved the quantity oriented water managers in active and passive water quality management. This means that they not only had to construct and to operate waste water treatment plants (active quality management), but also to work on the prevention of surface water pollution by means of permits and charges for waste water discharges (passive quality management). Water demand control and water quality protection thus became an additional focus of water managers (Grijns and Wisserhof, 1992).

The 1968 transition with respect to water demand control forms part of a transformation process which had already started in the 1950s. After World War II and a period of economic recovery in the 1950s, concerns were raised about how to meet the demands for natural resources (water, space, nature), needed by a growing economy and a growing population, which was also demanding a higher living standard. In that context, a Groundwater Act for Water Supply Companies was passed in 1954 to better guarantee a constant and undisturbed water supply. In 1962 a Physical Planning Act was adopted, allowing expropriation of land (for instance to the benefit of water drainage as a public service), and introducing disadvantage compensation for the effects of public planning on private property. In 1963 a Fisheries Act was introduced to prevent over-catching and to generate a more efficient fisheries industry. The Clearances Act [Ontgrondingwet] of 1965 and the Nature Conservation Act of 1967 were the start of the protection of nature and landscape resources against rival water and land uses. As a result, the 1950s and 1960s are characterized by controlled use expansion, a debate on public versus private interests, and redistribution of property rights (Van Hall, 1992). This was reflected in the first national water plan of 1968, since it mainly focused on how to meet the future demands of an increasing population and how to prevent rivalries related to water resources. The plan clearly recognized problems of groundwater scarcity and the need for demand-side management (Snijdelaar, 1993).

The 1969 transition with respect to water quality protection is also part of a transformation process which had already started in the 1950s. It has been a transformation from sanitation to quality protection. Sanitation was a focus of water management since the early 20th century, when sewage and supply systems started to be constructed as public services, mainly at municipal scale. After World War II the infrastructure for sewage and water supply gradually expanded into rural areas. Meanwhile the awareness was growing that sanitation measures would not be enough, and that something should be done about the increasing pollution of water resources, related to the expansion of economic activities after World War II. In the 1950s and early 1960s, many water boards and municipalities became active in the preliminary construction of waste water treatment plants, to prevent direct discharges of untreated sewage into surface waters. However, water boards and municipalities were insufficiently equipped to handle this in a systematic way. Initiatives were mostly restricted to areas with a high concentration of inhabitants and industrial activities. This changed around 1970, when many European countries, on the basis of international agreements, adopted legislation to protect surface water quality, which the Netherlands did in 1969. The content of such legislation varied greatly among European countries at that time. The Dutch Surface Water Pollution Act includes a system of permits to regulate industrial waste water discharges and a charge system with strong incentives based on the polluter-pays principle as well as the principle of full cost recovery. The costs of construction, operation and maintenance of waste water treatment plants was to be fully recovered from the polluters, and equivalent to the amount of pollution they emitted. The charge system applies to both industrial and domestic polluters (Leemhuis-Stout, 1992).

Although the institutionalization of water quality management has been an important step towards integrated water management, it did not involve a direct integration between water quality and quantity management. Initially water quality management was established as a sectoral water policy, with its own separate water planning. In 1975, 1980 and 1985 the Dutch environmental department introduced sectoral plans for water quality protection, while in 1968 and 1984 sectoral policy documents for water quantity issues were presented by the water department (Snijdelaar, 1993). While the Ministry for the Environment coordinated water quality issues, other water issues were controlled by the Ministry of Transport, Public Works and Water Management. In other words, the period from 1969 until at least 1985 could also be characterized as a period of increasing complexity and fragmentation.

3.2.2 The 1985 transition

A second important transition occurred around 1985 when the Dutch national water ministry issued a policy discussion paper on 'integrated water management' as a new approach for water managers. This advocated considering water as a system in which surface water and ground water are interconnected. The new approach would be not only to integrate quantity and quality aspects of the water system, but to also take account of the system's ecology. In fact, the second transition has been a crucial step in implementing a water basin approach which allows ecological considerations to enter water management decision making (Grijns and Wisserhof, 1992). The policy vision of 1985 was formalized as the Third Integral Water Policy Plan of 1988, prefaced by the Second Sectoral Water Policy Plan of 1984. The second policy plan recognized water depletion due to over-drainage as a major problem for water management. It also advocated integration between surface water and groundwater quantity management, but did not include quality aspects (Snijdelaar, 1993). Generally speaking, the 1985 transition was triggered by a general growing environmental and ecological awareness in society, as well as by a deregulation and integration tendency in politics in the early 1980s, resulting in a political demand for more coordination by means of policy planning. The same developments are clearly visible in Dutch environmental politics of the 1980s, culminating in the National Environmental Policy Plan of 1989, which was the first Dutch policy plan that intended to coordinate all sectoral environmental policies at the national level. Another important change in the 1980s was the Constitutional revision of 1983, which proclaimed that the public domain should be dedicated to the protection and sustainable improvement of the living environment, including the natural water system. The newly added Article 21 of the Constitution provided a fundamental title for expropriation of all property rights which could harm the protection of the living environment. Besides these more general triggers, there ware several more specific contexts in the water policy field in which the 1985 transition is embedded and which we explain further below.

After the extension of Dutch water management into the field of quality protection of surface water in the 1960s and 1970s, the 1980s became an important decade for the institutionalization of groundwater management in the Netherlands. In 1981 a Groundwater Act was adopted to further regulate extractions of groundwater, not just extractions by water supply companies, like the previous Groundwater Act of 1954. While the Act of 1954 aimed to better serve public supplies, the Act of 1981 intended more to redistribute extraction rights among all users with extractions above 100,000 m3, by creating a concession system. An important change introduced by the Groundwater Act of 1981 is that it proclaimed that the interest of public supply could no longer dominate the deliberation of interests. The Groundwater Act of 1981 should be considered an act that deals with water distribution, and not as an act that deals with water level management to prevent water depletion. For that goal, another act has been adopted in 1989, called the Water Management Act. This act introduced instruments for the level control of surface and ground waters, which could restrict all use rights affecting water tables, not only to prevent water depletion, but also to protect ecosystems. In fact, the Water Management Act has been the vehicle for the regional water boards, on the basis of which they could also use their ordinances for the regulation of water uses to protect the natural and ecological values of water systems (IJff, 1993; Teeuwen et al., 1993). It has been decided that ecological considerations are in the interest of the general public, and therefore the general public should have seats and be represented on the water boards. Since the charging system administered by the water boards is based on a 'profit-payment-participation' principle, the extended approach also implies that citizens should bear a certain share of the total costs of water management.

With respect to groundwater, not only its quantity aspects started to be regulated in the 1980s, but also its quality. In 1986, a Soil Protection Act was adopted to prevent, limit, and remediate changes in soil properties, especially applied to prevent agricultural contamination of ground and surface water. This resulted in the limitation of farming practice rights to protect groundwater, by means of standards for the application of animal and artificial fertilizers on soils. During the 1990s, the restriction of farming practice rights has gradually become more stringent, importantly induced by the EU Nitrate Directive for Groundwater.

Due to the rising influence of ecological and environmental considerations in Dutch water management during the 1980s, the need for interpolicy cooperation between water policy, environmental policy, nature conservation and agricultural policy increased greatly during that period. While sectoral policy plans for water, environment, nature and land use have become more integrated on their own during the 1980s, the integration between these policy sectors appeared to be very difficult. For instance, this has been very visible in the case of groundwater protection plans which are formally dealt with as part of environmental policy planning rather than water management planning, from which they are excluded for reasons of demarcation competency.

3.2.3 The 1995 transition

In 1995 the Dutch authorities fundamentally changed their traditional approach to river management and flood protection based on construction of dyke fortifications along the river banks. Triggered by serious river floods in 1993 and 1995, ascribed to climate change, a new policy document (entitled 'Space for Water') was presented in 1995, which stressed the need to integrate water management with land use planning. The document advocated the better anticipation at climate change by creating more space along river banks for water retention as a means for natural flood protection, in contrast to artificial protection by means of dyke fortifications. The Netherlands should be regarded as a delta area in a European setting. This new approach fits well with the water basin approach of 1985, since retention areas are a chance for natural and ecological restoration of water systems. Moreover, retention areas could help to replenish groundwater stocks in order to prevent water depletion in dry periods. So in fact, the 1995 transition has not only been triggered by river floods in the early 1990s, but also by the international alarm over climate change, the problem pressure of water depletion, as well as the preference to restore the natural flow of rivers and valuable ecosystems along the river banks. Like the 1985 transition, the policy document of 1995 was formalized by a Fourth Integral Water Policy Plan in 1998. Compared to the previous policy plan of 1988, this plan focused especially on climate change and on the restoration of the natural dynamics of water systems. It advocated regarding water and its natural movements as the key determining factor in spatial planning. It also emphasized the value of water in terms of open, unspoiled landscape (Hofstra, 1999). In 2000, a state commission on Water Management in the 21st Century recommended creating extra titles for expropriation of flood plains and for limitation of land use rights in flood plains. The commission also recommended dividing the liability for flood damage among the national water authority, regional water authorities and private property owners. Water boards could avoid such liability by developing sufficient areas for water retention, based on a system of safety standards which determines what the storage capacity in a region should be (Commissie WB21).

Although the 1990s have witnessed many attempts at integral river management and development of sustainable river basins, the policy sectors of water management and spatial planning are still rather separate. This renders decision making difficult, especially at the level of the water boards and municipalities, where the former have an interest in considering water as a guiding principle in spatial planning and leaving areas unbuilt if a risk of inundation exists, while the latter have a final say in spatial planning and have an interest in economic and urban expansion within their geographical boundaries. The coherence between water management and land use planning might increase as soon as the intended titles for expropriation of flood plains and for limitation of land use rights in flood plains go into effect.

3.3 The IJsselmeer: nurturing nature in an artificial lake

The IJsselmeer, with its size of 2,000 square kilometers and 600 kilometers of shore-line, is one of the largest freshwater lakes on the European continent (see Figure 3.1). For the greater part of modern history the lake was actually a sea that was directly connected to the North Sea and regularly plagued the Dutch shores. This sea was tamed in the first decades of the 20th century with the help of a 30-kilometer long closure dam ('afsluitdijk' - see map), which created the current artificial lake. The lake was named after the IJssel, which is the Northern branch of the Rhine and the river that feeds the lake.

Figure 3.1: *The IJsselmeer area. The different shades of gray in the lake indicate depth. At the deepest parts, this is 7 meters below NAP (the Amsterdam datum level).*

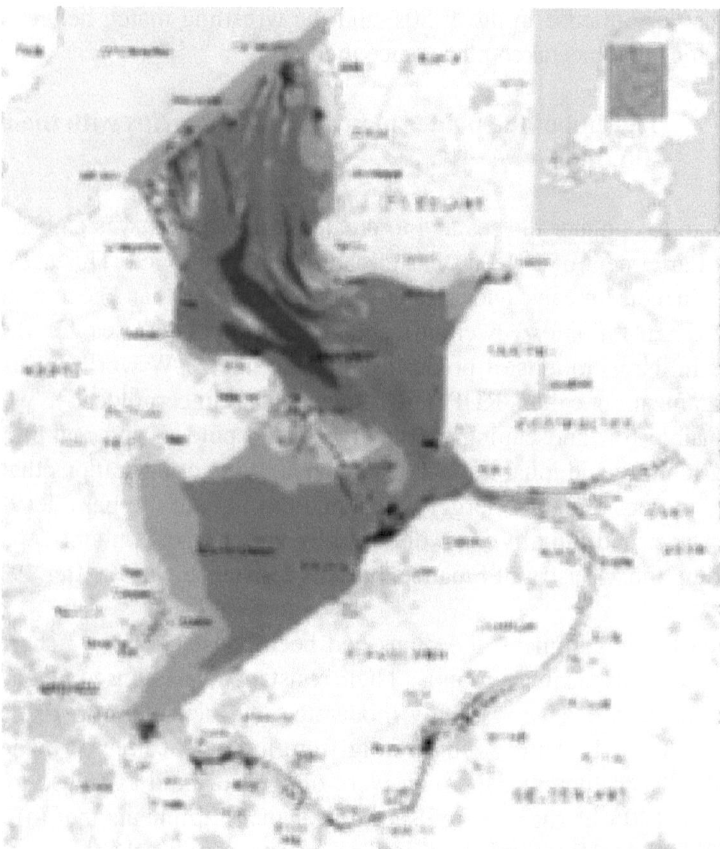

The plan to 'tame' the sea was so ambitious that it actually took a few decades to devise it, then another few decades to gain political acceptance, and then again decades to actually implement it. Besides a closure dam, the plan also included the creation of new land by turning parts of the lake into 'polders', to be used for agriculture and to increase safety by shortening the shoreline. Resistance to the plans had been great. Improved technology, food shortages in World War I, and an enormous flood in 1916 pushed aside concerns. From 1918 until well into the 1980s, subsequent Dutch governments remained committed to the project. However, from the early 1970s onwards, the plans that had been approved in 1918 started being seriously questioned. They were de facto abandoned by the government after fifteen years of public debate. Subsection 3.3.1 describes how this came about. Rather than being perceived as 'not yet reclaimed land', the lake came to be increasingly appreciated as a natural area with great value. In separate subsections we discuss two rivalries that have emerged since then: the rivalry between gas drilling and drinking water exploitation (3.3.2), and the rivalry between nature and recreation (3.3.3). Important as background to all these rivalries, however, is the discussion on the poldering plans, which became very intense in the 1980s, and the wrestling match between various authorities that has taken place since then.

3.3.1 The end of the polder plans. Coming to grips with the new situation

The plan to conquer the sea and to create the IJsselmeer was embedded in an act of Parliament of 1918, as has already been mentioned. This act furnished both financial means and created an organization that became known as ZZW (Dienst Zuiderzeewerken) and it controlled the area for decades to come. In its operations of poldering the lake, the ZZW worked closely with an organization called RIJP ('Rijksdienst IJsselmeerpolders'), which was responsible for landscaping the newly created polders and making them fit for agriculture and inhabitation. The agenda of the organizations that worked on the project remained largely the same throughout the period 1918-1970. The project had initially been opposed by various government departments, including the Ministry of Finance, but this resistance waned after 1918.

Outside the government, the plans had been resisted by the fishermen that used the sea as fishing grounds. Their resistance was consequential as their economic significance was only moderate, and their claim to fishing rights was rejected. The latter is interesting from the perspective of ownership of the lake. Even though fishermen were obliged to hold a license to fish the sea since 1905, Parliament refused to hear their arguments that (a) they had vested rights to fish, and (b) that these rights prohibited the project. In the view of Parliament, the sea was collective property under the control of the state. The fact that the fishermen would suffer from the project was considered a ‘normal entrepreneurial risk’. The act of 1918 did contain a promise to the fishermen that they would be helped to overcome the damage to their operations, but little financial compensation. Fishermen were offered re-education and also fishing rights in the newly created lake, which appeared useless at that time. The profound meaning of the 1918 act and the deliberations in parliament was that public ownership of the lake was established.

It was only after the creation of the closure dam and the first polders that the first interest in the natural and recreational value of the IJsselmeer came about. Working with ZZW and RIJP in a committee, environmental and recreational interests started influencing the actual landscaping of the polders. Instead of a long, straight coastline, the newer polders became more curved in shape, which was more interesting to see from sailing boats. Small islands for recreation were created. The gradient of the coast became less steep so that a richer bird life would become possible.

The ‘second’ wave of environmentalists, which arrived in the 1960s, was in many ways much more radical than the first wave. They took relatively little interest in the debate on the new polders, but this changed when the last polder (Markerwaard) that was yet to be constructed started being mentioned as a potential site for a second national airport. This was in itself related to a change in leadership at the ministry that oversaw ZZW, the Ministry of Transport and Public Works. The new Minister was an adept at social cost benefit analysis and had demanded a review of the polder plans. This decision had pressed ZZW in the direction of mentioning various possible beneficial uses of the last polder, among them an airport. Although ZZW succeeded in constructing the image of an economically worthwhile undertaking in its cost benefit analysis report, the agency lost precious time. In 1974, the government of the day announced an ambitious push towards national land use plans in 1974. The belief in planning was at its height, as was the discontent with the unstructured way land use decisions were being taken at the time. The Cabinet decided to introduce land use plans at a national scale, to be developed in highly participatory fashion. Added to this decision was a list of issues for which national plans had to be made. Among them was the plan to construct the Markerwaard polder and build an airport on it. The decision of the Cabinet had grave consequences for ZZW, in the sense that the water managers did not control ‘their’ project anymore, that the poldering project had to be discussed in land use terms, and that opponents now had a ready pathway to official decision procedures.

Even though the airport and the polder discussions were decoupled fairly soon, the polder discussion under the land use planning regime continued, with fierce criticism from opponents about the way ZZW advanced its course. The Cabinet of 1980 initiated the final phase of the discussion with a decision to actually construct the polder. However, the resolve to actually implement this discussion lessened quickly as time progressed and finally dissolved in 1991. The main driver for this decline of resolution was a fear of the enormous costs related to the project. A group of private parties made a bid to construct the polder in the mid-1980s, but this group withdrew when government insisted on a payment for the land 'under the lake'. Even though the 1991 cabinet decision formally ended the desire to polder the lake, the old reflex of planning development 'in the lake' (in practice this means that smaller parts of the lake need to be poldered) is still very much alive. For example, the city of Amsterdam recently built an urban area (40,000 houses) in a reclaimed part of the lake.

In the 1980s, with ZZW's grip on control loosening and the population in the polders rapidly increasing, a discussion started on future government system in the IJsselmeer area. The central government decided to divide the lake and polders over three provinces (one new: Flevoland) and multiple municipalities (six new). With the advent of 'ordinary' governmental bodies in the IJsselmeer, the importance of ZZW and RIJP, which later merged into a new organization called RDIJ, was considerably reduced. Disagreement erupted between RDIJ (or more broadly speaking, central government) and the local authorities over various issues such as the possible construction of a nuclear reactor along the IJsselmeer coast, gas drilling, the location of new recreation areas, and the appointment of nature areas. The provinces broke off consultations with RDIJ and started developing their own 'interprovincial plan' for the area, which conflicts in various respects with national priorities. The land use planning system is very decentralized, however, and therefore local authorities cannot really be forced to take over national priorities.

It is somewhat ironic that the water managers in the IJsselmeer (RDIJ but also certain water boards) nowadays have a much broader agenda than they had in the past. RDIJ is still very much concerned with flood safety and views many developments (i.e. climate change) in that perspective, but the agency has also become the water quality manager and has internalized a certain concern for the ecosystems in the lake, and until today RDIJ has retained its perspective on the entire water basin. All of this would have implied a broadening of the regime in the IJsselmeer, if RDIJ had not been dependent on the 'new' actors that entered the policy arena. The exact implications of the presence of these new actors are not necessarily negative from a sustainability perspective, however, as these would depend in part on the intentions of the new actors. We can assess this question especially in the sphere of rivalry between nature and recreation, and so we zoom in on this rivalry.

3.3.2 Gas drilling and drinking water exploitation

The soil under the IJsselmeer is quite likely to contain certain gas reserves at a few kilometres depth. We will not go into the legal details here (see Taverne, 1993), but do wish to point out that under Dutch law, the state owns the gas as it is owner of the IJsselmeer. Gas drilling is a staged activity in the sense that reserves have to be proven first, which is done by seismic tests and exploratory drilling. Government permission is required for such exploratory drilling. The government has chosen to exclusively allot 'search areas' to interested parties (see Figure 3.2), but even the parties that 'own' these search areas (mainly the companies NAM and Elf in the IJsselmeer) may not drill at a specific location without consent from central government. Nor does the actual discovery of gas reserves imply a right to actually exploit the reserves, but generally speaking it is the case that the company that proves the existence of certain reserves will be allowed to exploit them by government contract. On average, the state collects 85% of the proceeds, which can add up to considerable sums.

Drilling for gas in the Northern part of the IJsselmeer became an especially contentious issue in the 1980s. A set of test drilling sites had passed the central bureaucracy after two years of proceedings and their selection had not resulted in any public response at all. However, shortly before the Minister of Economic Affairs was about to sign the license, the provincial water company of Noord Holland (PWN) started making noises about the potential negative effects of gas drilling on its intake activities. The company extracts drinking water from the IJsselmeer and was worried that in case of accidents, certain quantities of oil could end up in the lake, thereby spoiling the water quality (Gerits, 1990). The likelihood of this event was extremely small, but the perception of the company (which had already permanently closed all other water inlets because of water quality concerns and which had just invested millions in water quality improvement measures) was that the risk was too grave. The company was backed by the provinces and most municipalities in the area and they started a fierce lobbying campaign to stop the license. This campaign was effective. The Minister did not sign the license, thereby leaving the company in question (NAM) in limbo. The company, however, is partly owned by the state and partly owned by certain large oil companies, Shell among them, which was rather cautious about conflicts with environmental groups.

Figure 3.2: *Drilling licenses granted in the IJsselmeer area. The interrupted lines indicate the outside borders of license areas; shaded areas are exempt from drilling activities. 'Boorvergunning' = drilling license.*

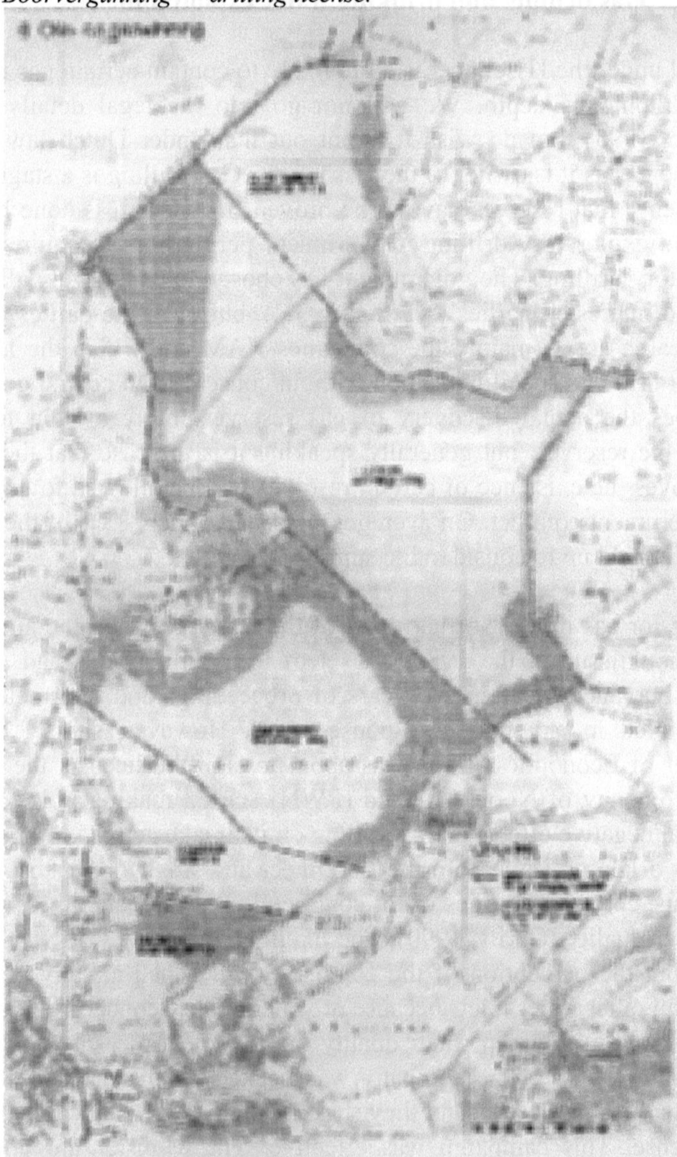

Since then, the situation has been in deadlock, with NAM and the Ministry of Economic Affairs silently waiting for the real possibility of a political mood swing, and the municipalities and provinces attempting to prohibit gas drilling through land use plans. The courts frustrated the latter attempts, however, as they confirmed that land use plans cannot alter existing drilling rights. It is not unlikely that the current surpluses on the Dutch government budget will evaporate again at some time. That is likely to be the moment when Parliament will become more sensitive to the prospect of increased gas earnings from the IJsselmeer.

3.3.3 The rivalry between nature and recreation

Both the recreational and the natural value of the IJsselmeer (and certain parts of the polders) came as a surprise. The IJsselmeer holds numerous attractions for tourists. The lake itself is the largest open space in the Netherlands, with the greatest undisturbed visibility. Around 17,000 boats (mainly sailing boats) have their home in the region, served by 128 harbors. The villages around the lake, nature reserves, and fishing grounds are the main destination for much daytime recreation. The total amount of spending associated with tourism in the area was estimated at two billion guilders annually. Table 3.1 below gives some other data.

Table 3.1*: Recreation in the IJsselmeer in the 1990s (adapted from RLG, 1999).*

Recreation form	Number of days spent at/ near the corner lakes (per year)	Number of days spent at/near the rest of the IJsselmeer (per year)
Tanning and swimming	250,000	100,000
Sailing and wind surfing	450,000	740,000
Motor yachts	2,600,000	130,000
Motor yachts (outgoing)	890,000	50,000
Rowing and canoeing	375,000	150,000
Angling	900,000	250,000

Even if we restrict ourselves to the lake and the immediate coastline, the ecological value of the lake is enormous. The basis for the ecosystem is algae growth, mussels (driehoeksmosselen), and to a lesser extent certain types of water plants. The importance of the Friesian coast has long been recognized and large parts of the coast have been assigned a status as natural reserve. There are many wading birds in the shallow parts of the lake, but many migrating birds use the lake as well. All in all, there are 25 bird species in such concentrations in the IJsselmeer as to make their habitat qualify as a Special Area of Protection (SPA) under the Bird Directive. Tourists come to the IJsselmeer for several reasons, but the presence of open space, an attractive coast, and nature are very important. Natural interests, of course, can also benefit from an association with recreational interest because it generates greater support for their existence and implies a connection with a politically and economically powerful sector. Despite the presence of such mutually beneficial relations, the possibility of discussions about the exact location of recreational and natural areas and the exact rules that regulate recreation remains very real, for instance because certain types of nature are rather sensitive to disturbance. By the mid 1970s, ZZW concluded that the enormous growth in recreation since the Second World War had done considerable damage to the natural beauty of the IJsselmeer. A round of 'integral plans', on an informal basis, started in 1977 for the ten so-called corner lakes (the bodies of water between the 'old' and the 'new' land). In a period of ten years, four integral plans were produced, but only one plan was formally endorsed by all participating parties. In most corner lakes, local authorities resisted limitations to tourist development. When the polders and the lake were formally incorporated into ordinary government structures, the voluntary cooperation between the government levels stopped. RDIJ continues to be concerned about developments in the area, such as the large number of illegal recreational settlements in natural areas (IIVR, 2000: 23). 'Administrative fragmentation' is an important cause according to the agency (IIVR homepage).

A range of different responses is being developed to address the situation. In the corner lakes, RDIJ has continued to play an important role, 'helped' by ecological disaster. The ecosystem of the corner lakes collapsed completely in the 1970s because of exponential algae growth related to phosphate and nitrate emissions. Under the powerful influence of RDIJ, an emergency plan was developed in 1986, which had resulted in very positive effects by 1995. Partly encouraged by this precedent, RDIJ started a new 'integral planning' exercise for all corner lakes in 1996, with the participation of about 20 government organizations and much public involvement. The outcomes of this decision process had just become available when our empirical study ended, but seemed positive in the sense that much coordination had taken place. Ratification of the outcomes by other actors was uncertain, however, as the plans actually implied a reduction of bird habitat and might not be acceptable under the Bird Directive. This was defended with a somewhat twisted line of reasoning: 'Model calculations' were said to have shown that 'the possibilities' for water fowl lessen by 5% because of the plan, but that "*this does not mean to say that the number of waterfowl will reduce by 5% (. . .)*". Furthermore: "*If previous actions within BOVAR [the plan to address algae growth, DH and SK] are ignored, the picture emerges that, on the whole, the foreseen measures do not fit in the desire to maintain existing bird values completely. However, when previous measures are taken into account, the picture changes. (...) the ambitions of the Bird Directive have been more than met, even before the Directive came into force"* (IIVR, 2000: 26). The reasoning is shaky at best. The Ministry of Agriculture, Fisheries and Nature Management was especially vocal in its criticism.

The second route towards coordination is along the path of land use planning. The three provinces surrounding the lake have worked together quite well and have already produced two interprovincial policy plans. The provinces devote much attention to tourism and developed a policy of concentrated growth, which they claim to be consistent with national land use plans. The latter is probably incorrect because the national government considers the IJsselmeer as 'the' place for growth in recreational activities that are to be phased out elsewhere, and the provinces are clearly more conservative in their endorsement of tourism. Interesting in that respect is the fact that especially the province of Noord-Holland has succeeded in achieving a high level of coordination among its municipalities along the coast, mainly by suggesting that they should develop a common vision and by threatening to use its oversight powers. The municipalities succeeded in achieving a common vision but it must be noted that one or two municipalities withdrew from the process because they were unsatisfied with their share of the pie. Since then, the province has refused approval of several municipal land use plans, which then resulted in litigation. So far it seems that the courts have upheld the provincial policies.

The third route towards more integration seems to be the private initiative. Especially the work in Noord-Holland has generated spin-offs in that respect. The Association for Water Recreation was retained by the municipalities to advise them on possible recreational development in the area. This Association is developing relatively unique databases on tourism, and has since then moved in the direction of vision development for the entire IJsselmeer. The Association has linked up with various other organizations, among them nature groups, and together they have developed their own plan for the lake, which will play an as yet unknown role in future policy development. It is interesting, though, that a few environmental organizations declined an invitation to talk with the Association for Water Recreation. It was suggested to us that this had to do with their expectation that they need not compromise their interests, given the presence of the Bird Directive. From the Dutch perspective, then, 'integral plan development' seems to be threatened by the Directive.

3.4 The Regge: undoing water management of the past

The Regge is a rain river in the Eastern part of the Netherlands, close to the German border (See Figure 3.4). The basin of the Regge is characterized by a mild slope, downwards to the northwest. The river basin is 900 square kilometres large and is bordered to the east and west by lateral moraines, caused by past glacier activity. From these moraines, which can attain heights of 30 meters above sea level, groundwater streams move downwards to the river valley. The area of the Regge contains about 500,000 inhabitants, concentrated in the built-up area (including the larger towns of Enschede, Hengelo, and Almelo). The area outside the towns is mainly rural with much agricultural land, but it also contains some important nature reserves (forests, peat land). Land use in the Regge basin is divided over agriculture (61%), natural areas (24%), and built up area (15%).

Figure 3.4: *River basin of the Regge. Taken from Van der Vlist (1995). The dotted lines mark the border of the basin. Indicated are the towns in the area (shaded), groundwater protection areas (black lines), and the various streams that constitute the Regge and feeding streams (dark lines).*

The manager of the Regge basin is the water board ('waterschap') Regge en Dinkel and this organization plays a central role in water management, together with the province of Overijssel and, somewhat more remotely, the Public Works Agency ('Rijkswaterstaat'). The activities of the board in many ways reflect the demand for water services by the users of the land in the Regge basin. Founded to combat flooding of agricultural land, the board propagates flood protection. The board also wants to make sufficient water for agricultural purposes available in summer. Both purposes are enhanced by a complex of measures that include canalization, maintenance of waterways, admitting water from elsewhere, and barrages in the Regge (one for every seven kilometres of river). This system is of great importance for the groundwater tables in the Regge area. The water board has statutory duties of water quality management, which it seeks to achieve by a license system and the operation of wastewater treatment works. Especially in the summer, the effluent from these treatment facilities constitutes a significant part of the amount of water that is in the riverbed, making water quality a concern. This concern is enhanced by the fact that, for the purpose of flood management, much of the cleaner water from the top of the basin has been redirected elsewhere (see Figure 3.5).

Figure 3.5: *disconnected parts of the basin. At least 60% of the basin has been disconnected. Arrows indicate the direction in which water is diverted.*

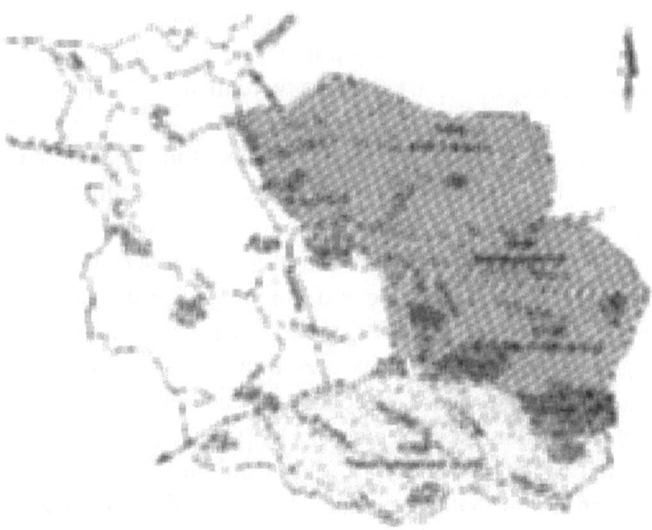

In line with its agricultural roots, the water board advocates efficient use of agricultural land, which was enhanced for decades by land relocation and the canalization of 90% of the water streams. More recently, the water board has started to regard itself increasingly as an organization concerned with the appearance of the landscape and with ecology. Moreover, flood protection is increasingly undertaken through the creation of storage capacity within the basin (retention) rather than increasing transport elsewhere. These changes in water management largely came about in the 1980s through a range of factors, including the demise of agricultural control over the water board council and changing national priorities. We focus on these changes in the remainder of this chapter.

3.4.1 The first broadening of the water regime in the Regge basin

What we would currently describe as damage to the landscape was long and widely regarded as improvement, signified by naming the activities 'improvement works'. Starting in the 1960s, water board intervention in the natural landscape caused conflicts with environmental groups but also with other layers of government. As happened in the IJsselmeer, the land use planning system was instrumental in channeling claims for greater attention to nature. The regional plan ('Streekplan') for Twente of 1965 is a case in point. A draft of this plan had been developed by the province of Overijssel and listed nature development as the top priority, implying restrictions on water board activity. The water board, left out at the consultation stage, objected to the procedure that had been followed and suggested that the regional plan was at odds with its own policies. After much consultation, a compromise was struck and various changes were made to the plan.

Despite that, the position of the water board had fundamentally changed from the only public body engaged in management of the countryside, to one of the players involved. The improvement works (and even regular maintenance) by the water board now required planning permission. Various groups sought to protect the landscape and fought against the canalization of various streams. Some municipalities were sensitive to these pleas and started limiting water board activity in natural areas. On various occasions, municipalities and the water board stood against each other in court. Donker (1996: 119-128) writes that the province of Overijssel used its responsibility for oversight of the water board to warn that it did have a responsibility for nature conservation and should complain less about developments in land use planning. The water board rejected this role, and in doing so, could refer to a 1967 national inter-departmental report, which advocated a narrow task conception for water boards. The development of nature or recreation areas

was explicitly mentioned as something that a water board should *not* engage in.
In 1974 the conflicts with environmental groups and other tiers of government inspired one member of the water board council to call for the establishment of a separate environmental commission, but such plans gained no majority whatever (ibid.: 127). The water board council did decide it would assign the task of consulting with the National Forestry Service ('Staatsbosbeheer') and environmental groups to one of the committees of the water board Council (the Finance Committee). This had certain effects on water management, although they cannot be described as very fundamental. The two most notable changes were the termination of the use of chemicals to fight weeds in the water and a more nature-friendly water bank maintenance.

3.4.2 Bornsebeken plan: a change of paradigm

Despite the changes in water management just described, the water board continued to work on its main agenda dating back from shortly after the Second World War. This agenda revolved around canalizing the last remaining streams, lowering the groundwater table and increasing transport capacity out of the Regge basin. In the 1970s, the water board started developing a so called 'structure plan' for a part of Regge called the Bornsebeek. This plan, appearing in 1978, demonstrated that improvement of the situation in the Bornsebeek would involve much work in the connected creeks as well. The Bornsebeek plan (single) became the Bornsebeken plan (plural). The project would essentially involve the entire set of creeks to the east of the Regge, at least the ones located between the towns of Enschede, Hengelo, Almelo, and Oldenzaal (10,000 hectares).By 1983, the water board itself concluded that the plan was too expensive and had disadvantages in terms of water quality and effects on the landscape. In accordance with emerging (national) philosophy of the 'broad outlook', the plan would be reconsidered. The original plan would have implied damage to the landscape as a consequence of the canalization (especially the trees along the creeks) and water quality in some creeks would worsen, because wastewater coming from the towns would be spread over all creeks instead of concentrated in one stream.

It took until 1986 before a working group of municipal, provincial and water board representatives, working with a consultancy firm, had developed a concrete plan for the Bornsebeken that seemed to meet the various concerns. The essence was that water would be diverted from certain streams to others. For that purpose, certain new connecting canals had to be dug and numerous creeks would have to be deepened and widened. Creeks used to be 5-7 meters wide, but would now be become 15-20 meters wide (ibid.: 11) Creeks with high natural value were to be spared, but this was not possible throughout the entire area. Other effects included a lowering of the groundwater table in agricultural land.

Subsidies and societal resistance largely determined the fate of the plan, in that order. Subsidies were an important factor in this sense that the national subsidy scheme for the improvement of regional waterways was suddenly withdrawn. The water board had seen the change coming, and had sped up development of the Bornsebeken plan, but it was not completed before termination of the subsidy scheme. The speed with which the Bornsebeken plan had been developed and its content led to an outcry from various organizations, including large landowners (Stichting Twickel) and environmental groups (e.g. Nature and Environment Federation Overijssel (NMO)). The province of Overijssel was critical as well. All these actors considered the project too one-sided.

The critique was taken to heart, it seems, and the working group that had developed the water board plan now established a so called 'alternatives committee'. The Committee identified seven possible alternatives for the original plan. Especially the ideas to allow flooding to occur and to create retention ponds were intensively discussed, with water board representatives voicing concerns over the costs and the amount of land needed for this solution. The committee's final decision was that the existing plan was better than all alternatives. The representative of the Ministry of Agriculture Nature Management and Fisheries, however, did not agree to the outcomes. At the Minister's request, a group of students wrote a report on the Bornsebeken plan. They developed an Integral Water Management Plan in Twente (abbreviated to PIWAT) that heavily criticized the water board plan as being 'mono-functional'. The PIWAT plan was different, in the sense that it pleaded for giving the river free reign ('meandering'), for retention areas, less drainage capacity of the waterways, acceptance of a higher level of flooding, and separation of water streams with good and poor quality (ideas that were already circulating in the context of the Third Note on Water Management). The response of the water board was that implementation of PIWAT would require the water board to step outside its legal responsibilities, which was something they did not intend to do. The PIWAT plan was rejected.

The Minister of Transport, Public Works and Water Management, however, was made aware of the PIWAT study too and she noted the obvious fact that new national policy directions were well embedded in the plan. In a Solomon's verdict, the Minister suggested she was willing to grant a subsidy for the water board's plan but that she was not convinced that it was integral enough. To guarantee that the improvement works would be implemented in an integral fashion, the Minister suggested that a new plan needed to be developed by all authorities involved. This group would operate under the supervision of the province, and consensus in the group would be a precondition for any subsidy.

A new working group started its activities near the end of 1990. The water board assembled an interdisciplinary working group that assisted them. The group started with an inventory of already existing ideas, particularly at the national level. The issue started to broaden because there was space to bring forward all problems that existed in the basin. The new project, for instance, was linked with the older recommendation to concentrate wastewater in certain streams, so that at least some other streams would achieve better quality. The water board had also received an indication that the water soil was seriously contaminated with mercury in certain parts. A solution to this problem (excavation of the water soil) became part of the project. Ideas from others were now also reconsidered. The inlet of water from the Twentekanaal was reconsidered and an opportunity was suddenly seen to reconnect the creeks south of the canal to the creeks north of it. The idea of a retention area was also taken on board (40 hectares of flooding area). Studies were commissioned on groundwater levels in the area, and the idea that groundwater tables should perhaps rise was discussed. By October 1991, the water board executive approved of a draft of the new plan and consultation with organizations outside the working groups started.

This time, nature organizations were much more positive, although it was noted that the plan still contained proposals for drainage and would not be sufficient to combat drought. From the agricultural side there was much criticism. According to the GLTO (Regional Agricultural Organization) the original plans had already paid sufficient attention to nature. Agriculture had suffered long enough from the flooding, and now the water board would take much of their land away. And that for a problem caused by urban development! The farmers feared most the expropriation of land, but also expropriation of use rights. Especially the fact that natural areas tended to have implications for nearby agricultural land was pointed out. To most of these questions the water board replied that land would only be taken on a voluntary basis and would be compensated by other land or by finance. The comments did not lead to many changes in the vision, and by May 1992, the authorities involved in the working group informally approved it. Procedures to change municipal zoning ordinances were started near the end of 1992. By 1993, the Minister personally came to the water board to endorse the plans. Much resistance from farmers has accompanied the land use planning procedures that have followed since. Although, in most cases, the water board succeeded in friendly negotiations with the various landowners, some have continued to object to the various stages of the plan.

3.4.3 Influence: a new way of working

The Bornsebeken plan has had a lasting impact on water board policies. New personnel were attracted to be able to approach the project in a multidisciplinary fashion. These people (biologists, ecologists, etc.) brought new visions of water management with them and could better connect to networks such as the environmental NGO's. New insights regarding water management were thus anchored in the (top of) the organization and these came to be reflected more generally in the water board policies. Many of the people involved with the Bornsebeken plan would later become involved with the development of a vision for the Regge basin as a whole (Regge Vision, 1996) and the similarities between the two plans are striking (separation of water streams, retention areas, etc.). Note that the initial hesitation of the water board about the PIWAT plan had been based on a narrow conception of the water board's task. Many of the goals of the PIWAT study could only be realized through the cooperation with other actors and the more conservative water board officials did not want to be responsible for that. After preparation of the Bornsebeken plan, the officials holding that opinion were largely gone, however. A new breed of officials entered the scene, with a taste for exactly the kind of interaction with other governments that had initially been rejected. At the national level, many of the new ideas on water management essentially implied a warp into the domain of other policy sectors such as land use planning. Regge en Dinkel got an early taste of that through the integral version of Bornsebeken plan.

3.5 What do the cases illustrate for the Netherlands?

In the previous section we have discussed developments towards integrated water management for two different water bodies in the Netherlands, the IJsselmeer and the Regge river. In both cases the extent and the coherence of the regime have considerably increased on many of the important aspects during the period 1970-2000. In case of the IJsselmeer in fact three rivalries have been discussed. First a rivalry about the use of a water body for land reclamation versus the value of open water. Secondly, a rivalry between the use of a water body for gas drilling versus the potential use of the water body for drinking water preparation. Third, a rivalry between the use of a water body for recreation versus natural values.[1] The presentation of the Regge case has added a fourth rivalry between the drainage function of a river basin for agriculture versus the desire to restore and further develop the ecological values of the water system. In this section we analyse for all four rivalries the regime changes that have been observed (both in terms of a changed extent and a changed coherence), the implications of the observed regime changes for sustainability, as well as the change agents and conditions by which the regime changes have been induced.

3.5.1 Regime changes in terms of extent

The extent of a regime refers to the extent to which relevant uses and users of a water basin are recognized by the regime. It is an indicator of the completeness of the domain of the regime. In both cases we have described, a considerable increase of the extent could be observed. In case of the IJsselmeer the extent started to increase around 1974, when the maintenance of the lake and various uses related to open water (except for ecological uses) came onto the government agenda. Open water related aims were uses that were not recognized before. Later on (around 1985), ecological values of the lake came also onto the agenda. However, in the 1990s plans were still made and implemented for small scale land reclamations (urban expansions outside the dikes close to Amsterdam, Lelystad and Almere). It took until 2000 before these plans were more generally considered as unfavourable in terms of water safety and management (by the national Commission on Water Management 21st Century, although other actors -- like the province of Flevoland -- still advocate the need to reclaim more land in the IJsselmeer for urban expansion). The extent also increased with respect to the rivalries around gas drilling and recreation. The potential negative effects of gas drilling on natural values (like birds) and on the quality of drinking water reservoirs in the IJsselmeer are fully recognized now. The IJsselmeer area is nowadays regarded as the 'Blue Heart' of the country, to be preserved. The values of nature and bird life in the corner lakes are recognized, although they are coming under pressure from increasing recreation settlements. In the case of the Regge river, the period 1989-1992 has been a demarcation. In

[1] In the final chapter of this book, in which all European cases of this book are analysed together, a fourth rivalry about fisheries in the IJsselmeer area is also taken into consideration. Different from the other rivalries between heterogeneous resources users, the fisheries sub-case is an example of a rivalry between homogeneous users. To keep this chapter on the Netherlands within the margins of an acceptable length, we have chosen not to describe and analyse the fisheries sub-case here.

that period the water board, which is the water management authority for this river basin, adopted a reconstruction plan for a subsystem of the Regge river (Bornsebekenplan). This was an attempt to serve both agriculture (drainage, flood protection) and the natural values of the water system. For the water board this has been a crucial step towards the adoption of an integral approach of water systems throughout its entire area.

3.5.2 Regime changes in terms of internal coherence of public governance

The coherence of a regime distinguishes between the internal coherence of public governance, the internal coherence of property rights, and the external coherence between public governance and property rights. First, we consider the internal coherence of public governance. This is the degree to which the interdependencies in the water system and its management are reflected in the elements of public governance. In this book we distinguish five elements of public governance: (1) levels and scales; (2) actors and networks; (3) perspectives and objectives; (4) strategies and instruments; (5) responsibilities and resources for implementation.

With respect to *levels and scales* we have seen that the state is the main water authority in the IJsselmeer basin, while a regional water board is the main water authority in the case of the Regge river. In both cases we could observe that increased interaction with other administrative levels has increased the complexity of water management, which in case of the IJsselmeer resulted in integration attempts accompanied by fragmentation, and which in the case of the Regge river resulted in successful integration attempts. With respect to the rivalry around land reclamation in the IJsselmeer, the state (as resource owner and main planning administration) and the provinces (as involved planning administrations) have emerged as relevant authorities. However, it remains unclear which one of them has the primacy in planning. Fragmentation could be observed since 1989. With respect to the rivalry around gas drilling in the IJsselmeer, the state (as water resource owner and as Ministry of Economic Affairs, implementing the Mine Act) decides on concessions for gas drillings and their locations. Emergence of the provinces since the 1980s, being responsible for the availability and quality of drinking water reservoirs, has led to high complexity and fragmentation. Together with the municipalities, the provinces have obtained land use planning responsibilities to ameliorate any negative effects of gas drilling. On the other hand, this fragmentation is in favour of more integrative decision making, because land use authorities at the provincial and municipal level are forced to deliberate between the interests of gas drilling and drinking water reservoir protection. Such an integrated deliberation of both interests is not provided by the national level of government. With respect to the rivalry around recreation in the IJsselmeer, since 1977 the Ministry of Transport, Public Works and Water Management took the initiative for integrated coordination at the scale of the corner lakes, including the participation of the provinces and municipalities involved. However, fragmentation at the national level could be observed between the Ministry of Transport, Public Works and Water Management, taking care of integral water management, and the Ministry of Agriculture, Nature Conservation and Fisheries (LNV), taking care of the implementation of the EU Bird Directive. In case of the Regge river, interventions by the state and the province were needed to force the water board, dominated by agricultural interests, towards a more integral approach. In the end, however, the water board proved to be able itself to adopt an integral plan at basin level. So integration at basin level has been successful.

With respect to *actors and networks*, in both cases we have seen strong indicators of increased access opportunities and participation by new users. In most cases a policy planning system has been the access channel for new users. With respect to the rivalry around land reclamation in the IJsselmeer, the discussion on the land-water conflict has been impelled by NGOs, such as the 'Vereniging tot Behoud van het IJsselmeer (VBIJ)', and also by recreational interest groups. With respect to the rivalry around gas drilling in the IJsselmeer, provinces and municipalities have become important players in the arena, often in coalition with environmental pressure groups which oppose gas drilling. With respect to the rivalry around recreation in the IJsselmeer, the Ministry of Transport, Public Works and Water Management organised participation for all interested parties. In the case of the Regge river, a working group has been assembled, supervised by the province, to prepare the integral plan. All the authorities involved participated. Consensus in the group was made a precondition for any subsidy. After a draft plan was approved, consultation with organizations outside the working group started (nature as well as farmers' organizations).

With respect to *perspectives and objectives*, we generally find a powerful tendency towards integration. In case of the rivalry around land reclamation in the IJsselmeer, the perspectives of different actors and user groups have converged. Only one actor (province of Flevoland) still sees the lake as 'not yet land'. Flevoland does not have the resources to polder, which they would like to do. The objectives of the water management sector have turned from a classical approach of engineering water works (until 1980) into an approach of protecting other water values and the importance of water areas as storage basins (as a response to climate change and rising sea level). In case of the rivalry around gas drilling in the IJsselmeer, there are several visions for the lake. The picture of this rivalry is more one of fragmentation.

'Official' government visions come both from the provinces and from the national government. The plans are at odds with each other. This situation is made possible by a desire to start land use planning in the lake. However, the EU Bird Directive and the national (drinking) water policy documents provide a perspective for the protection of birds and drinking water resources in the IJsselmeer. On the other hand, these documents do not really provide a single direction perspective to solve the rivalry. In case of the rivalry around recreation a round of integral plans started in 1977 for the corner lakes. In case of the Regge river, the Bornsebekenplan has become an integral vision (after the rejection of several previous versions), in which the perspectives of the rival users have converged. The approach developed has strongly encouraged the development of an integral vision on the river basin as a whole, which resulted in the Regge Vision in 1996.

With respect to *strategies and instruments*, we could observe that planning has been the most important policy instrument for adopting new uses in the policy process and for getting user groups around the table. In case of the rivalry around land reclamation in the IJsselmeer, the introduction of a new planning procedure (PKB) for large infrastructural works in 1974 has resulted in a better recognition of rival uses. However, at the end of the 1990s water management authorities still lacked adequate instruments to get more grip on land use planning. In case of the rivalry around gas drilling in the IJsselmeer, the main instruments seem to be (land use) plan formation and concession permits on basis of the Mine Act. These are at odds, and increasingly so. However, since 1989 an Environmental Impact Assessment is required for gas drilling. This is seen as an important instrument to achieve more integration. In case of the rivalry around recreation in the IJsselmeer and in case of the Regge river, plan formation is also the main instrument to achieve integration. In the Regge case, the supervisory role of the state and the province and their ability to facilitate with subsidies have been used to force the water board to develop an integral approach.

With respect to *responsibilities and resources for implementation*, we could observe that implementation still is fragmented in the IJsselmeer case, while it is much more concerted in the Regge case. With respect to the rivalry around land reclamation in the IJsselmeer, in the 1970s land use planning instruments had to be applied to get the value of open water adopted by the state water management sector with its one-sided focus. However, during the 1980s this sector changed from a rather closed into a much more open policy community, which very much broadened its scope and accepted the value of rival uses. Conversely, at the end of the 1990s we see that the water management sector has problems in getting the value of open water adopted as a guiding principle in land use planning. With respect to the rivalry around gas drilling in the IJsselmeer, responsibilities are fragmented and very little coordination is possible, which leads to conflict. The gas drilling issue goes to court many times. On the other hand, the EU Bird Directive is working as an influential tool or resource which strengthens the drinking water interest. In case of the rivalry around recreation in the IJsselmeer, the integral visions had to be implemented by the system of land use planning, and by that means they had to be adopted (ratified) by local ordinances. That didn't work very well. Besides that, formal responsibilities are not where the money is. The national government has the money; provinces and municipalities have land use planning authority. In case of the Regge river, the plan formation functioned as a concerted action for the implementation. All authorities involved have been committed to applying their resources. Support was built by consulting representatives of the main user groups. However, land had to be acquired to implement the plan. The water board declared it would take only land on a voluntary basis and compensate with other land or finances. Despite this statement, individual farmers still objected in procedures to change the municipal zoning ordinances. In most cases, the water board succeeded in friendly negotiations with the various landowners, although some others maintained their resistance.

3.5.3 Regime changes in terms of internal coherence of property rights

The internal coherence of property rights refers to the degree to which the interdependencies in the water system and its management are reflected in the distribution of property rights among the users involved. In the cases described, the internal coherence of property rights generally increased. With respect to the rivalry around land reclamation in the IJsselmeer, the state (holding the ownership of the IJsselmeer) has greatly restricted its own decision-making power to polder by committing itself to the outcome of a PKB-procedure, which attributes participation rights to rival users. This happened around 1974. In case of the rivalry around gas drilling in the IJsselmeer, the internal coherence became more complex and fragmented. The state ownership of the IJsselmeer and the use right by the state to allow concessions on gas drilling are in conflict with the use rights of provinces on drinking water reservoirs as well as with the regulatory rights of provinces and municipalities to help other users effectuate use rights which are threatened by gas drilling. An indicator of more integration in this case is that the obligation of an Environmental Impact Assessment for gas drillings is restricting the use rights for gas drilling. On the other hand, the courts seem to be uncertain about the extent to which these restrictions could hold

With respect to the rivalry around recreation in the IJsselmeer, municipalities and provinces initially had an influential say (by means of regulatory rights) in the siting of recreational settlements. Natural life in the corner lakes was hardly protected by property rights. However, recreation is increasingly brought under land use controls, although there is still much illegal land use. Nature rights are increasingly recognized by the land use planning system too. The land use planning system is improving the internal coherence, since this is a system which attributes by ordinance property rights to the uses involved. The EU Bird Directive is another change which could attribute property rights to bird life in the corner lakes. However, divergence (fragmentation) could be observed between the flexible nature protection provided by the land use planning system and the strict nature protection by the Bird Directive. In case of the Regge river, nature is hardly protected by property rights. However, nature rights are increasingly recognized since the water boards are allowed (by the Water Management Act of 1989) to promulgate ordinances which incorporate ecological considerations in decision making on water basins.

3.5.4 Regime changes in terms of external coherence between public governance and property rights

The external coherence between public governance and property rights is the degree to which changes in public governance are reflected in changes of the property rights. In the cases we described the external coherence has increased on only a few of the important aspects. With respect to the rivalry around land reclamation in the IJsselmeer, the right to polder (reclaim land) has been restricted in favour of access rights for rival users. Nowadays, the right to polder (in the hands of land use planners) is still out of the control of water managers wanting to protect open water. Although it is expected that this will change in the future (the state is working on legislation), it still has not been put into effect. Therefore, the external coherence has changed from high in the period 1974-1995 to incomplete in the 1990s for this rivalry. In case of the rivalry around gas drilling in the IJsselmeer, the status of property rights is still not clear. Many government actors would like to stop drilling, but this is difficult. The Environmental Impact Assessment obligation has increased the external coherence only somewhat in some of the more important aspects. In case of the rivalry around recreation, the property rights of municipalities according to the land use planning system are not sufficiently redistributed to guarantee nature protection rights, despite the protection goals of the Bird Directive and Dutch nature policy. In case of the Regge river and the implementation of the Bornsebekenplan, the water board had to rely on voluntary cooperation of landowners, as far as they are willing to provide land against compensation. In that respect, the change has been more policy driven than property rights driven. However, municipal zoning ordinances are used to restrict use rights on land that is needed for the natural development of the river basin (to attribute property rights to nature).

3.5.5 Implications for sustainability

If we want to assess the implications of the observed regime changes for the sustainability of the regime, we could draw a distinction between: (1) implications for natural resources and the environment; (2) implications for economic development; (3) implications for social development. We are only able to make a general assessment of implications of observed changes for sustainability. The overall sustainability of the resource use is beyond our capacity to judge as social scientists.

Considering *implications for natural resources and the environment*, we could observe that in the case of the IJsselmeer the uses related to open water have become much more widely recognized. After the introduction of the PKB procedure in 1974, there still have been cases of land reclamation in the IJsselmeer-area. These cases are disputed nowadays, and the importance of open water for safety (climate change, rising sea level) has become increasingly recognized since the mid 1990s. With respect to gas drilling, we have to notice that they still might affect the lake, especially if oil leaks occur. Until now, the concessions for gas drilling have not been activated and still can be reconsidered as a result of the rivalry debate. This is a bit of an artificial problem, since it is mainly a reflection of society's changed priorities. The fear of possible risks related to gas drilling has led to a rethinking and a court battle on the sustainability of gas drilling in the IJsselmeer. With respect to the recreation rivalry, we may conclude that nature is better protected than it was. Although the implementation of nature protection in the municipal zoning ordinances doesn't work well, the authorities try to channel tourism away from the sensitive/valuable areas. The strong use-driven policies to encourage recreation in the corner lakes (opportunities for tourist and economic development) prevented substantial reductions of recreation settlements. Most authorities that are involved, except for the Ministry of Agriculture, Nature Conservation and Fisheries (LNV), still try to run away from the implications of the Bird Directive. In case of the Regge river we could observe that natural and ecological values of the river basin are nowadays much better recognized and expressed in water management by the water board.

Considering *implications for economic development*, we could observe that in case of the rivalry around land reclamation in the IJsselmeer, the loss of water has been capitalized and incorporated into cost-benefit analyses of land reclamation projects. With respect to the rivalry around gas drilling, it is clear that the option of not drilling would cost the economy quantities of potential income due to the government. However, the drinking water reservoirs have been valued more highly and will strongly effect the cost-benefit analysis of gas drilling. In case of the recreation rivalry, we notice that significant investments in waste water treatment have been done to fight against growth of algae and to improve the life conditions in the corner lakes. In case of the Regge river, the water board is spending part of its taxation revenues for nature development. The expanded extent has been accompanied by the entrance of new participants (inhabitants and representatives of nature organisations) in the water board, which since then have started to contribute with taxes to the water management tasks in which they have an interest. Important provincial and national subsidies are also involved.

Considering *implications for social development*, we observe that the decision making on land reclamation (poldering) has become the subject of an open debate with the public. Gas drilling activities have become the subject of an intensive public debate in society, with great political sensitivity. This could help the protection of drinking water reservoirs. In case of the recreation rivalry, the value of the lake for bird life went much higher up the agenda, especially as a consequence of the EU Bird Directive. In case of the Regge river, the water board has gone through an existential debate on its mission. The nationally introduced 'broad water system approach' of 1985 has been finally accepted by the farmers which were rather dominating the water board.

3.5.6 Agents for regime change

Looking at the agents, which in combination have forced regime change, we make a distinction between: (1) national regime changes; (2) EU policies; (3) problem pressure; (4) other change agents. We are only able to consider the combined force of the listed change agents as an impetus to set in motion regime changes in the direction of more integration.

In case of the rivalry around land reclamation in the IJsselmeer, we observed three *national regime changes* that have driven regime change at basin level. First, the adoption of a Physical Planning Act in 1965 has been influential, together with the changing political climate in the 1970s and growing political legitimacy of groups protesting against the establishment in general, and in particular against the classical approach of water management (engineering) by Rijkswaterstaat. A second impulse has been the incorporation of ecological values around 1985. A third impulse has been the national attempts to integrate between water management and land use planning in the period 1998-2000. In terms of *problem pressure*, first the gradual loss of natural values in reclaimed water areas has been a change agent. Second, an increasing awareness of insufficient water safety in the 1990s, in particular the river floods of 1993 and 1995 and related flood damage liability concerns, have triggered regime change. A third change agent has been the alarming report in 2000 by the national Commission on Water Management 21st Century, demanding serious attention for problems of climate change, rising sea level and water safety in the country. *Other change agents* have been first a decreasing need for new land and a gradual decline of the agricultural sector. Second, alternative options for the Markerwaard, especially the plan for a new national airport in that area, has led to growing protest against expansion of urban areas and infrastructure at the cost of natural areas and open space. Third, budget reforms in the early 1980s have definitely changed the willingness of the state to invest in land reclamation.

In case of the rivalry around gas drilling in the IJsselmeer, we observed two *national regime changes* that have impelled regime change at basin level. First, the national recognition of groundwater scarcity and increasing drinking water needs (around 1988-1989: Third National Water Policy Document and Policy Document on Supply of Drinking Water and Industrial Water) has led to awareness of the importance of drinking water reservoirs in the IJsselmeer. It has also triggered the provinces to look eagerly at these reservoirs. Second, the introduction of more stringent requirements regarding environmental impact assessment (around 1989) has been a change agent. As far as *EU policies* could trigger regime change, it is relevant to notice that the IJsselmeer has become a Special Protection Area (SPA) under the EU Bird Directive as of 2000. However, until now the directive hasn't been applied in this case to help protect the drinking water reservoirs. In terms of *problem pressure*, it is uncertain to what extent drilling would actually affect drinking water quality. Probably there is only a small risk. However, if the gas market opens up, there will be a more powerful incentive to sell and explore for gas, and the awareness of this has increased the fear of problem pressure in the near future.

In case of the rivalry around recreation in the IJsselmeer, the integral water system approach, introduced by the Ministry of Transport, Public Works and Water Management in 1985, has had a major impact as a *national regime change* on the adoption of nature protection values and ecological considerations in the land use planning system. *EU policies* have not yet become a change agent, but could become a greater force in the future. The strengthening of natural values by the Habitat and Bird Directives is important. The status of the IJsselmeer as a Special Protection

Area potentially limits tourism. In terms of *problem pressure*, environmental degradation of the corner lakes (with enormous growth of algae as an indicator) has triggered much more active work on nature protection. On the other hand, chances for tourist and economic development prevented substantial reductions of recreation settlements. The national Ministry of Agriculture, Nature Conservation and Fisheries (LNV) even promotes the transfer of recreation from the Wadden Sea in the north and from the Province of Zeeland in the south to the IJsselmeer area in the heart of the country.

In case of the Regge river, the integral water system approach, introduced by the Ministry of Transport, Public Works and Water Management in 1985, has also been the most influential *national regime change*, which triggered the adoption of nature protection values and ecological considerations. In terms of *problem pressure*, social resistance (by nature and environmental organisations, as well as country-seat owners) have largely determined the change of the traditional drainage approach into a more integral approach. In terms of *other change agents*, we could observe that subsidy programs have had a powerful impact on regime change at basin level.

3.5.7 Conditions for regime change

For making a general assessment of conditions that have been favourable or unfavourable for regime changes in the direction of more integration, we distinguish the following conditions: (1) tradition; (2) joint problem; (3) joint opportunities; (4) credible alternative threat; (5) institutional interfaces. The conditions should be regarded together to assess their influence.

In case of the rivalry around land reclamation in the IJsselmeer, *tradition* has been a favourable condition in the sense that earlier cases of land reclamation in other polders and the Lauwerszee (in the Province of Groningen in the north of the country) have triggered the state to think more of incorporating natural values and the value of open water into water management. At that time, too, the awareness of environmental and natural values was growing. Protest groups were hotly disputing land reclamation by the state water authority (Rijkswaterstaat), while advocating a more open decision making process on water works and claiming more participation rights. Initially, the perspective of a *joint problem* was absent. There wasn't much common knowledge in the form of reports, or information symmetry between the actors. However, among politicians, overlooking these disputes, a sense of responsibility for the future value of open water was growing rather quickly. So a joint problem turned out to be a favourable condition in the end. Another favourable condition has been the existence of *joint opportunities*. There were lots of protest groups demonstrating against land reclamation for various reasons. However, they shared a common interest in stopping the classical land reclamation process by Rijkswaterstaat. A *credible alternative threat* on the side of the state and national politicians has been that they could meet the strong wish of protest groups to skip the Markerwaard land reclamation, by using budget reform considerations as a formal argument for it. The alternative option (land reclamation) would have had much more severe budgetary consequences than the choice for integration. An important *institutional interface* has been the PKB planning procedure for major infrastructural works, which was introduced in 1974 on basis of the Physical Planning Act, and which provided a legal leeway for a more integrated approach.

In case of the rivalry around gas drilling in the IJsselmeer, a *tradition* of integrated thinking has been absent. The absence even of a *joint problem* has been an unfavourable condition. Very different visions of the problem resulted in a heterogeneous understanding of the problem. Distrust exists between national bodies and the provinces, as far as this rivalry is concerned. *Joint opportunities* are lacking, since there is no information symmetry and no notion of possible joint gains from integration. The provinces and municipalities involved don't want to respect the almost established rights to drill. A *credible alternative threat* is not available. The central government cannot, or rather will not, compel the provinces because of the political costs. There is a legal path for integration through the land use planning system, but political opportunity forbids this use. However, in terms of *institutional interfaces*, national water policy documents are in favour of protecting the drinking water reservoirs in the IJsselmeer. This has very much motivated the provinces to oppose the gas drilling. The Environmental Impact Assessment obligation and the Bird Directive could be seen as a legal leeway for more integrated approaches.

In case of the rivalry around recreation in the IJsselmeer, a strong *tradition* of integrated thinking at the national level has been a favourable condition. However, this tradition wasn't that strong at the local level, so it was not shared by all actors. In terms of a *joint problem*, there is a certain awareness among the rival users that nature is the basis for recreation, and must be preserved. Some municipalities oppose tourism and are happy to let others develop it. This should also be seen as a *joint opportunity*. Since the presence of attractive nature is an attractive condition for tourism, and since tourism is a way of experiencing nature, there has been a sense of respect for each other's interests. In terms of a *credible alternative threat*, no imbalance of power could be observed. The national government (also a resource owner) is strongly advocating nature protection and integral solutions, but they are very dependent on the local zoning ordinances. The municipalities do not really have a strong interest in integration. The provinces could compel the municipalities, but they also have an interest in tourism. The Ministry of Agriculture, Nature Conservation and Fisheries (LNV), responsible for the implementation of the EU Bird Directive, could become an important player, but isn't now, as far as this rivalry is concerned. In terms of *institutional interfaces*, the approach by the Ministry of Transport, Public Works and Water Management of forming steering committees to stimulate integral visions and to work on agreements among the actors involved, has been a favourable condition.

In case of the Regge river, a strong *tradition* of consensus and mutual respect, characteristic of the democratic functioning of water boards, has been a favourable condition for a change of attitude on the farmers' side. The absence of a *joint problem* was initially an unfavourable condition. Very different visions of the problem existed, resulting in a divergent understanding of the problem. There was also distrust between farmers and non-farmers. However, it has been a *joint opportunity* that the alternative, advocated by the opponents of natural development, didn't harm the interest of the farmers in getting river flood problems solved. It was only that they didn't want their money to go to nature development and to sell agricultural land for it. In terms of a *credible alternative threat,* the dependence on subsidy programs has been in the joint interest of all parties involved to develop a common plan, as well as being an instrument by which the national and provincial authorities could force change. In terms of *institutional interfaces*, the nationally introduced 'integral water system approach' as well as the increased societal resistance against traditional drainage solutions (canalization) have forced an integral solution. The national and provincial authorities could use their supervisory role (and their subsidy resources) to force through an integral plan.

3.6 Conclusion

One of the objectives in this book is to evaluate institutional regimes at water basin scale from the perspective of the EU Water Framework Directive. This directive aims to make water management at basin scale more integrated, and through that more sustainable, by expanding the scope of water management, by getting citizens involved in water management, by streamlining legislation for water management, and by getting the prices right for water management services.

In the Dutch cases of this chapter we have seen that the *scope of water management* at basin scale has changed from a rather mono-functional and economically driven use orientation (land reclamation, gas drilling, recreation, drainage for agriculture) into a multi-functional perspective which not only focuses on resource use but also on resource protection (sustainable use). The recognition of new values, like open water, drinking water for the future, and natural resilience, has increased the extent of water management. In fact, we have shown that the increase of extent has been driven by a combination of national regime changes, EU policies, problem pressure and a few other change agents. As far as national regime changes are concerned, the adoption of rival values by physical planning around 1974 as well as the national regime transitions of 1985 and 1995 have been important for the recognition of natural values and the resilience value (natural flood prevention) of water basins. As far as EU policies are concerned, the Habitat and Bird Directives are becoming increasingly important for the protection of natural values, but are not fully implemented and used in that way. As far as problem pressure is concerned, societal resistance to mono-functional water management, the gradual loss of natural values, and the increasing awareness of insufficient water security (due to river floods in the 1990s) have been of great influence. As far as other change agents are concerned, the gradual decline of the agricultural sector and the economic appraisal of natural values (for instance by means of subsidy programs) have been important triggers, inter alia.

The increased extent has been accompanied by an increased *participation of new users*. In addition to the traditional mono-functional users of water basins, which were already strongly represented in water management, citizens have become more involved in water management, especially as representatives of the new values and the common interest of sustainable resource use. Planning has been the most important instrument for providing them access, as we have seen in the case of land reclamation in the IJsselmeer around 1974, and in the other cases in the 1980s. Access rights in planning procedures have been formalized at the national level by the adoption of various legal provisions which aim at a broader deliberation of interests in water management. Access rights for citizens in the regional water boards have also been formalized at the national level. The increased participation of new users initially resulted in more complexity and fragmentation of water resource regimes due to rivalries between the different users that want to claim a specific use of the resource. In fact, the water basin regimes changed from a simple (mono-functional) regime into a complex (multifunctional) regime. A positive implication of the increased complexity has been that the 'new uses' went higher up the political agenda and that the 'old uses' have become the subject of an open debate with the public. However, in all cases it has been difficult to achieve an extra step from fragmentation to integration, and thus from a complex regime to an integrated one.

The way to integration is not only a matter of improving the internal coherence of public governance, but also a matter of improving the internal coherence of property rights, as well as a matter of improving the external coherence between public governance and property rights. With respect to the *internal coherence of public governance*, we have seen that the water basin *scale* has become more central in both cases. In case of the IJsselmeer, the provinces are playing a stronger role in addition to the traditional control by the state. In case of the Regge, the regional water board has gradually developed a total plan for the water basin as a whole. Looking at *actors and networks*, we have observed that participation of new users has increased. Looking at *perspectives and objectives*, we noticed that the extent has increased. Looking *at strategies and instruments*, we concluded that planning has been an important instrument for integration. In terms of *implementation*, we observed much more concerted action in the Regge case than in the case of the IJsselmeer. With respect to the *internal coherence of property rights*, we have seen that in the cases of land reclamation and recreation in the IJsselmeer the deliberation of rival uses has increasingly been brought under land use controls. In the case of gas drilling, use rights are restricted by the obligation of an environmental impact assessment. In the case of the Regge, regional water boards are allowed to incorporate ecological considerations in their ordinances for water level control. Although these changes of property rights have increased the internal coherence, they are only changes in the sense that a careful deliberation of interests is required. They do not guarantee that the new uses may rely on a property right on their own. The only exceptions are the EU Habitat and Bird Directive, which really do attribute property rights to natural values, but which still are not put into effect in that way in the Netherlands. The consequences for the *external coherence between public governance and property rights* are that the Dutch consensual approach of deliberating interests by means of planning instruments does not always guarantee a proper protection of new uses, and therefore does not guarantee a high external coherence. For instance, in the case of land reclamation in the IJsselmeer it is still difficult to provide sufficient protection to the value of open water and water security. Even the requirement introduced (at the end of the 1990s) that local land use plans always need to be assessed by water boards in terms of water risks, is only a guarantee for interpolicy cooperation between municipalities and water boards, but not a redistribution of property rights. So the integration tendency in the Netherlands has mainly been a public governance driven tendency, and much less a property rights driven tendency.

How can one evaluate the Dutch situation with respect to the wish of the European Union to *streamline legislation* for water management at basin level? It is clear that the Dutch legislation has been streamlined to facilitate a better deliberation of rival uses. The Dutch streamlining has been rather weak in the sense of redistributing property rights among rival users. Until now, the Dutch have been very reserved in treating the EU Habitat and Bird Directives as a tool for redistribution of property rights. Another EU criterion is to *get the prices right* for water management services. Evaluating the Dutch situation from that perspective, we observed at least clear attempts to capitalize the value of new uses and to incorporate them into cost-benefit analyses. This has especially been the case for the value of open water and the value of drinking water reservoirs. In the Regge case we have seen that taxes are paid by citizens to generate revenues for nature development, as well as that subsidy programs have become available for developing a more natural water basin.

Considering the conditions that have been most important for more integrated water management at basin level, we may conclude that, speaking generally for the Dutch situation, the tradition of integrated thinking, which started in the 1960s, and the much older tradition of consensus and mutual respect have been influential in the evolution of water basin regimes towards more complexity and integration. Although problem perspectives in the cases started to diverge widely, and therefore the perspective of a joint problem mostly has been absent as a condition, in some cases the presence of joint opportunities or a credible alternative threat has been important as a condition for achieving a breakthrough. In all cases, institutional interfaces have been a most favorable condition, in the sense that national policy documents and national initiatives have been an important trigger for change at water basin level. This conclusion supports the image of the Netherlands as a so-called 'decentralized unitary state', in which a strong tendency towards subsidiarity and decentral autonomy exists, as long as it fits within a nationally harmonized approach. The history of Dutch water management has shown that national interventions in regional and local water management always followed after events that threatened the water safety of the country at a supra-regional scale. Although the Dutch have started to anticipate climate change through water management at a basin scale and by water boards as basin authorities, it is to be expected that the national authorities will again intervene powerfully after future calamities.

REFERENCES

Commissie WB21 (2000) *Waterbeleid voor de 21e eeuw*. Den Haag.

Disco, C. (red.) (1998) Waterstaat. In: J.W. Schot e.a. (red.) *Techniek in Nederland in de Twintigste Eeuw. Deel I.* Zutphen: Walburg Pers, pp. 52-207.

Donker, H. (1996) *Water tussen Regge en Dinkel. Waterschapszorg in Twente tussen 1934-1984*, Almelo.

Gerits, R.T.F. (1990) *Milieueffecten van gasboringen in het IJsselmeer*, Amsterdam.

Goverde, H.J.M. (1987) *Macht over de Markerruimte*, Nijmegen.

Grijns, L.C., J. Wisserhof (1992) *Ontwikkelingen in Integraal Waterbeheer. Verkenning van beleid, beheer en onderzoek.* Delft: Delft University Press.

Hall, A. van (1992) Naar een samenhangend waterbeheer, bezien vanuit de kwantiteitszorg. In: S.B. Boelens e.a. (red.) *Waterstaatswetgeving. Verleden, heden en toekomst.* Bundel ter gelegenheid van het honderdjarig bestaan van de Staatscommissie voor de Waterstaatswetgeving. Zwolle: W.E.J. Tjeenk Willink, pp. 155-182.

Hofstra, M.A. (1999) De Vierde nota waterhuishouding en de Vijfde nota ruimtelijke ordening. Verbanden in planvorming en integratie van beleid. In: A. van Hall e.a. (red.) *De Staat van Water. Opstellen over juridische, technische, financiele en politiek-bestuurlijke aspecten van waterbeheer.* Lelystad: Koninklijke Vermande, pp. 77-89.

IIVR (2000)(Integrale Inrichting Veluwerandmeren), Inrichtingsplan Veluwerandmeren. Schakel tusen strategie en uitvoering (draft of 10 October), Lelystad.

IJff, J. (1993) Omwentelingen in het waterschapsbestel 1968-1993. In: J.C.N. Raadschelders en Th.A.J. Toonen (red.) *Waterschappen in Nederland. Een bestuurskundige verkenning van de institutionele ontwikkeling.* Hilversum: Verloren, pp. 13-29.

Leemhuis-Stout, J.M. (1992) Twintig jaren (regionaal) waterkwaliteitsbeheer. In: S.B. Boelens e.a. (red.) *Waterstaatswetgeving. Verleden, heden en toekomst.* Bundel ter gelegenheid van het honderdjarig bestaan van de Staatscommissie voor de Waterstaatswetgeving. Zwolle: W.E.J. Tjeenk Willink, pp. 183-196.

Regge Vision (Regge Visie) (1998), Almelo: Waterschap Regge en Dinkel.

Snijdelaar, M. (1993) Ontwikkelingen in de waterstaatszorg vanaf de jaren vijftig. In: M. Snijdelaar e.a., *De waterstaatszorg in Nederland. Verankerd in het verleden, flexibel naar de toekomst.* Den Haag: VUGA, pp. 9-35.

Taverne, B.G. (1993) *Beginselen en voorbeelden van regelgeving en beleid ten aanzien van de opsporing en winning van aardolie en gas*, Delft.

Teeuwen, H.H.A. e.a. (1993) Omgaan met de brede kijk door waterschappen. Gaan we breed keuren? In: M. Snijdelaar e.a., *De waterstaatszorg in Nederland. Verankerd in het verleden, flexibel naar de toekomst.* Den Haag: VUGA, pp. 37-54.

Vlist, G. van der (1995) *Duurzaamheid als planningsopgave. Gebiedsgerichte afstemming tussen de ruimtelijke ordening, het waterbeleid en het waterhuishoudkundige beleid voor het landelijke gebied,* Wageningen.

Chapter 4

Diverging Regimes within a Recently Federalised State

The Vesdre and the Dender in Belgium

David Aubin and Frédéric Varone
Université catholique de Louvain (Belgium)

4.1 Introduction

Belgium is a wet country that paradoxically has few water resources at its disposal, considering the high population density. Furthermore, the country faces major regional disparities. Flanders is dependent on Wallonia for 60% of its drinking water needs. There are a series of recurrent problems with water management: regular floods in many places due to major rainfalls in wintertime and serious problems with the pollution of surface and groundwater. A combination of a high population density, a long industrial tradition and intensive agriculture are increasing the pressure on the resource. The significant demand for drinking water leads to total exploitation of the aquifers. Nevertheless, improvements are being made in the fight against pollution. New uses and concerns appear with the development of leisure activities. The population is more aware of environmental concerns and is developing a greater consideration for nature. In this context potential rivalries between heterogeneous groups of uses and users are to be expected. The selected stories of local water management in Belgium reflect these problems and evolution.

Water management in Belgium is developing in a context of broad institutional change. From the beginning of the 1970s the former central State started to evolve progressively towards federalism. Three Regions are recognised, viz. Wallonia, Flanders and Brussels-Capital, which are entitled to an increasing degree of policy responsibility. The Belgian State became a

Hans Bressers and Stefan Kuks (eds.), Integrated governance and water basin management. Conditions for regime change towards sustainability, 99-129.

fully federal state in 1993. In the field of water, most competencies have been transferred to the Regions, in 1980 for the quality aspects and 1990 for the quantity aspects. This situation has led to different regional water regimes in Flanders and Wallonia.[1] The choice of the case studies has been motivated by a desire to compare the evolution of the two main regional water regimes in Flanders and in Wallonia. The cases are tributary river basins corresponding to units of management in the implementation of the EU Water Framework Directive (WFD). The Vesdre river basin in Wallonia and the Dender river basin in Flanders are representative of the major water problems in the Belgian Regions. The Vesdre case considers the issues of drinking water and mineral water production, the pollution of rivers and floods. The Dender case considers the dependence of Flanders for drinking water supply, and also the issues of surface water pollution, floods and nature conservation. In each case we present several rivalries between heterogeneous groups of users and the mechanisms leading to their resolution. The stories we tell occurred between 1980 and 2001.

The analysis of local water management is conducted around two interrelated research questions. Which conditions allowed the resolution of rivalries between heterogeneous water users? Do these rivalries and conflict resolutions go towards more sustainable water uses, one purpose of the WFD?[2] The chapter presents stories of the resolution of rivalries in tributary basins (sections 4.3 and 4.4). Both stories and basins are compared (section 4.5) and placed in the larger context of the institutional water regime in which they occur (section 4.2).

4.2 Deepening and divergence of the regional water regimes

Initially Belgium was a unitary country, created in 1830. Since it belonged to the French Empire, the country inherited of the Civil Code of 1804. This private law still governs property rights (i.e. formal ownership rights, disposition rights and use rights) including those to water. Public policies on water started to develop from the second part of the 19th century. The central objective of public health supplanted the priority accorded to the

[1] The case of the Region Brussels-Capital is not presented here. The Region Brussels-Capital received autonomy only in 1989 and its territory is limited to an urban area. Thus the problems faced by Brussels are traditional to such areas, i.e. a huge demand for drinking water, provided by plants located outside the Region, here in Wallonia, and discharges of domestic wastewater affecting another Region, i.e. Flanders. In the last case, Brussels is investing heavily in sewage treatment works.

[2] Art. 1.b. of the Directive 2000/60/EC of the European Parliament and of the Council of 23 October 2000 establishing a framework for Community action in the field of water policy.

development of agriculture and industry since 1893. The objective of environmental preservation appeared later, in the 1970s. The main public laws adopted before federalisation are the laws of 1950 and 1971 on the protection of (surface) water against pollution. The law of 1971 attempts to organise water management at a river basin scale (i.e. Meuse basin, Scheldt basin and Coastal basin). The designed basins cross the Regions. In the context of federalisation, this law which fits with the logic of the new WFD, has failed in its implementation.

The emerging process of Belgian federalisation put a stop to the attempt of the 1971 law, opening the way to regional responsibility for water in 1980. The public domain, and particularly navigable rivers, was transferred from the central State to the Regions, as were most environmental policy responsibilities. Since then, the Regions have managed their water independently. This situation, provoked by institutional reforms (external to the water policy), has led to diverging regional water regimes, all of which nevertheless lead towards more complexity and integration. Each Region has had to set up its own administration to manage these questions. Flanders has implemented the law of 1971 on the protection of surface water against pollution and has completed the framework legislation. Wallonia preferred to restart with new legislation, more suitable to the existing actors in place, i.e. the inter-communal water associations (*intercommunales)*.

4.2.1 Protection of drinking water sources and ecosystems in Flanders

After 1980, Flanders started to systematise water protection while embarking on a policy of hydrous independence. The institutional water regime aims at protecting wells from (diffuse) pollution, regulating discharges through global permits and developing their own capacities to produce drinking water. A relative water scarcity and a situation of dependence on Wallonia underlie these regional orientations. The Flemish Region has implemented the ('federal') laws of 1971 on the protection of surface and groundwater and has adopted environmental permits for industrial discharges. Rights to withdraw groundwater are limited by a procedure of prior authorisation, as for every hazardous activity. The Region levies fees on industrial emissions and taxes on households in order to finance wastewater treatment. The Region has confirmed its leading role in water policy with a complete review of the former institutional arrangement. As a result, regional water companies have been created, e.g. the *Vlaamse Maatschappij voor Watervoorziening* (VMW), i.e. the Flemish water distribution company in 1983, and the *Vlaamse Maatschappij voor Waterzuivering* (VMZ), which operates the sewage activities for the whole Region. The regional administration manages the rivers, controls withdrawals and discharges, and monitors the quality and quantity of water. The *Administratie Waterwegen and Zeewezen* (AWZ) -- the national Ministry of Public Works prior to 1990 -- governs navigable rivers, including dikes and floods, while the Administration of the Environment, *Administratie Milieu-, Natuur-, Land- and Waterbeheer* (AMINAL), administers smaller rivers and groundwater.

In the early 1990s the Region started to intensify wastewater treatment, regulate discharges by means of global permits, limit manure disposal and define absolute protection zones. The goal of the regime is both to improve the reserves of potential drinking water and to preserve ecosystems and biodiversity. Formal ownership rights to land, including wetlands and riverbanks, have been amended for purposes such as nature conservation and the development of recreational activities along the rivers. Expropriation has been used as an effective regulation tool. During the same period, the management of the quality of surface water has been reinforced. As a reaction to the persistent pollution of rivers and to European obligations, Flanders has partly privatised wastewater treatment and extended the regulation of (direct and indirect) discharges to new categories of users, particularly the farmers with measures limiting fertiliser application. The *Vlaamse Milieumaatschappij* (VMM), a public company, monitors the quality of surface water and sets down the investment plans for sanitation. These plans are accepted by AMINAL and implemented by Aquafin NV, a private company, that builds and operates the collective treatment plants throughout the regional territory. Aquafin concentrates the knowledge and expertise concerning sanitation. Despite the process of partial integration, mechanisms of co-ordination with the other water uses are still absent. Consultation structures at a local level, i.e. the basin committees (11 *bekkencomité's* for the whole Region), do not fill the gap.

4.2.2 Preservation of water quality in Wallonia

The picture is quite different in Wallonia, where the water resource is more abundant and related to more economic interests, such as tourism. The Region has hindered the development of an environmental policy. It did not implement the ('federal') laws of 1971, waiting until 1985 to regulate the protection of surface water and 1990 for groundwater. The Region protects wells, regulates discharges and plans the restoration of streams in order to preserve the quality of waters intended for the abstraction of drinking water and the attractiveness of rivers for tourists. Limitations on disposition and use rights are used as a means to conserve nature and limit pressure on the aquifers. Raw surface water or groundwater intake for drinking water supply needs a prior permission. As in Flanders, the Region takes the leading role, but it leaves more room to the local authorities; communes are closely involved through the inter-communal associations of supply and sanitation. Most competencies on water belong to regional administration, mainly to the *Ministère de l'Equipement et des Transports* (MET) for the management of navigable rivers and hydraulics and to the Administration of the Environment, the *Direction générale des Ressources naturelles et de l'Environnement* (DGRNE) for all other aspects.

The motivations for regime change in Wallonia are the same as those in Flanders, i.e. a persistent pollution of water and the need to satisfy the European requirements. At the end of the 1990s, Wallonia chose to reinforce water protection measures with a process of contracting between the regional authority and the industrial water operators. It also developed the regulation of discharges and the protection of specific areas through global permits, all in order to preserve the quality of raw water intended for drinking water supply. As in Flanders, the major innovation concerns the actors' network. The *Société publique de Gestion de l'Eau* (SPGE), a public company, was set up in 1999. According to a management contract signed with the Region, it manages all financial affairs linked with wastewater treatment and the protection of wells. The entire water supply and sanitation sector is integrated in a coherent framework supervised by the SPGE. The problem is that co-ordination between the water distribution cycle and other uses are not considered, e.g. minimum flows or nature conservation. Marginal, local mechanisms are developed, i.e. the river contracts (14 *contrats de rivières* in Wallonia), that work on the voluntary basis of the actors involved.

4.2.3 Towards cross-regional regimes?

When comparing the development of the regional water regimes, we observe, on the one hand, that the Flemish regime is confident in its amendment of ownership rights in favour of nature protection. It has also integrated the management of surface water quality and privatised its treatment facilities. On the other hand, the Walloon regime only intervenes on use rights for nature preservation and has integrated the water supply and sanitation sector with contracting and financing mechanisms. As a basin approach is emerging, the remaining problem in Belgium is that the river basins cross the Regions. As a result of the federalisation process, the Regions decided to manage economic and environmental matters on their own. It was thus not likely that true basin authorities would be created, as initially required by the 1971 law. However, cooperation between the

Belgian Regions at the scale of the river basin can nowadays be expected in the context of the International Commissions of the Scheldt and the Meuse. River basin management is on the way but it still remains an institutional arrangement that yet has to be set up.

Thus it makes sense to select two case studies in Flanders and Wallonia to compare the diverging institutional water regimes that might influence the resolution of rivalries and the sustainability of the resource.

4.3 The Vesdre river basin in Wallonia

After a brief description of the tributary basin of the Vesdre and a localisation of the problem pressures in the basin, we describe stories of rivalries that occurred during the period 1980-2001. The resolutions of rivalries mobilised local users and different elements of the regional regime.

4.3.1 Description of the basin

The Vesdre basin is located in the north-eastern part of Wallonia. It occupies a surface area of 710 km^2 and has a population of 210,000. The administrative boundary of the basin fits with the catchment area. The river Vesdre has its source inside the basin and flows into the Meuse. Its main tributary, the Hoëgne, comes from the south and flows downstream to Verviers. The basin receives no incoming water from outside, except for significant rainfall that influence the water flow. The relief inside the basin is mountainous and landscapes are varied. Nearly half of the territory is covered by forests. Housing and industry are concentrated along the river Vesdre, particularly between Verviers and Liège. Farming activities are located in the North. These are extensive and have a low impact on the quality of water. The river Vesdre is particularly polluted, with industrial and domestic discharges being the main pollution sources. Its industrial past caused damage and concentrated the population along a narrow perimeter (see Figure 4.1). Today, the river is lined with abandoned industrial areas.

Sewage has been a preoccupation since the 1950s. Long networks of sewers were built along the streams, but only moved the pollution. In fact, they were not connected to treatment plants. The pressure of wastewater discharges is concentrated downstream of Verviers, not just because it is the major agglomeration: it is also the place where the sewers discharge water directly into the Vesdre. Wastewater from Eupen is discharged downstream of Verviers, just as the water from Spa. Nowadays treatment plants are in operation or under construction at the mouth of these main sewers.

Figure 4.1: *Identification of rivalries in the Vesdre basin, 2001*

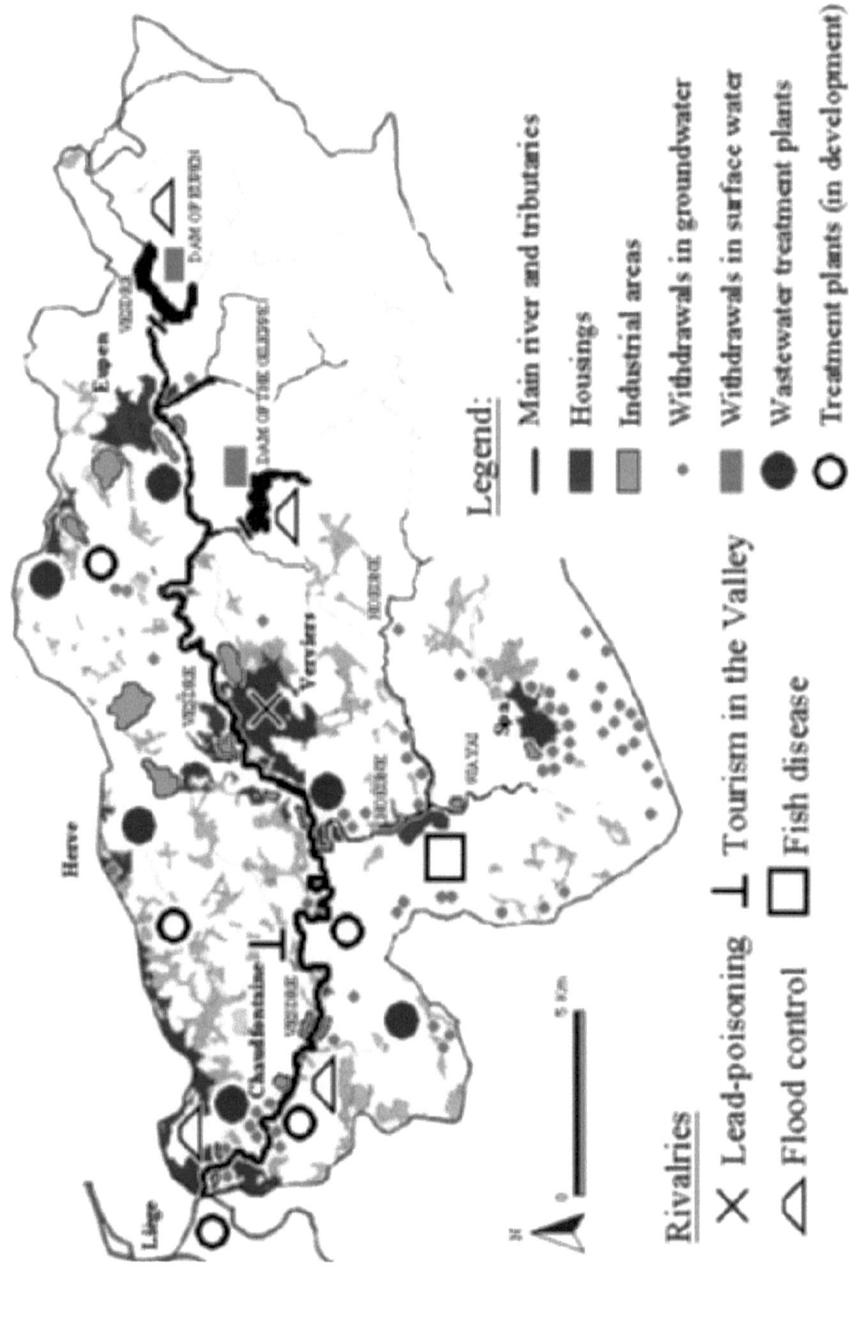

Source: background from DGRNE, data from DGRNE and own data collection

Concerning water supply, most industrial and drinking water is provided by the dams of Eupen and the Gileppe. The rest is provided from numerous aquifers. Groundwater wells are located along the river Vesdre or in the surroundings of Spa.

The local rivalries identified in the Vesdre basin are located at different points inside the basin. The first rivalry, concerning the lead-poisoning of tap water, is localised on the town of Verviers. Water comes from the Gileppe dam and is distributed in the town for industry and housing. The second problem concerns water floods in the lower part of the Vesdre (Chaudfontaine and the suburbs of Liège). In this case, the plaintiffs are not located in the same place as the defendant, the dams, located in the upper valleys of the Vesdre. The third rivalry, pollution of surface water, occurs all along the Vesdre, mainly between Verviers and Liège where most of the water discharges were displaced. Finally, the fourth rivalry, industrial pollution in the Wayai, occurs downstream of Spa, but is enlarged to all the water users of the tributary basin of the Hoëgne-Wayai. Thus rivalries occur in many places and not systematically between neighbours.

4.3.2 Lead-poisoning in Verviers

In the town of Verviers, a localised rivalry between the distribution of drinking water and industrial water gave rise to a conflict in the mid-1980s. Citizens of Verviers opposed the municipality (as water supplier), claiming that tap water should conform to European standards. Raw water comes from the dam of the Gileppe and is not treated. Moreover, this water is naturally acid and attacks the lead-pipes. As a consequence, water consumption in Verviers has led to lead-poisoning for more than a century. The same water supplied industry. Industrialists find the acidity of the water beneficial to their processes, due to its cleansing properties, e.g. for washing pipes, wool and paper. They did not want any change to this peculiar quality. Thus it was necessary to reconcile drinking water and industrial supplies.

The problem of lead-poisoning appeared in the late 19th century and has been well known since then. Initially the provision of a booming wool industry was enhanced thanks to the construction of the dam in 1878. The right of withdrawal from the dam reservoir belonged to the commune of Verviers, which distributed water to industry. The population soon demanded a connection of houses to the raw water mains, which the municipality implemented. Drinking water was not treated. For years the status quo persisted because of diverging interests and a poor knowledge of the nature of the contamination. The recognition of lead-poisoning automatically involved financing of a drinking water treatment plant.

In the 1970s concrete scientific proof of the contamination was established. The first project for a drinking water treatment plant was planned in 1977, but postponed many times due to competition between national administrations and later the federalisation process. The municipality could not finance such a plant. Moreover, industrialists were reluctant to a change in the acidity of the water and exerted pressure on the municipal council.

The conflict itself appeared between the municipality and an association of local citizens at the beginning of the 1980s. The citizens were annoyed by the poor quality of the water, not only about the well-known problem of lead-poisoning, but also about the colour and taste of the water. They decided to sue the municipality in 1984, their action being based on the European directive on the quality of drinking water (European Communities, 1980). Belgium adopted the directive in 1984, but did so badly: it included a specific exemption on lead concentration for Verviers. The association won the case on appeal in 1987. In parallel, it also referred the matter to the European Commission, a claim that ended up with a condemnation by the EU Court of Justice in 1990 for the failure of Belgium to implement the directive.[3]

As a result, the municipality committed itself to introducing modifications. A temporary solution was provided by the *Entreprise régionale des Grands Transports d'Eau* (ERPE), being part of the Walloon Region. It modified the source of supply as connecting the drinking water mains to the Eupen dam. However, this arrangement only covered 80% of the population. The situation was left unchanged for 10,000 people in the poorest part of the town until the necessary drinking water treatment plant should enter into operation. At the same time, the ERPE was building the plant, which finally came into operation in 1992. Concerning the complaint of the industrialists, a technical arrangement safeguards their supply of raw (and acid) water.

Ultimately, the public health concern was taken into account without inducing any redistribution to the detriment of other water uses. Many triggers were necessary to reach a solution. Quality standards on drinking water were defined and legally binding. On the one hand, an organised group of actors knew the rules and activated it, leading to an enforceable court decision. On the other hand, the regional authorities (ERPE) took charge of the construction of the drinking water treatment plant and found a solution to the claim of the industrialists. The case law was supplemented by the financial intervention of the State (here the Region) in a context of strong pressure by the people affected by the problem.

[3] Case law C-42/89 of 5 July 1990.

4.3.3 Chronic floods in the lower Vesdre

Water floods in the lower part of the Vesdre create a rivalry between industrial water uses, drinking water and protection against floods. The rivalry is caused by the two dams of Eupen and the Gileppe, which retain water upstream in the basin. The dams were built to satisfy growing needs for industrial and drinking water. In case of major winter rainfall, when the dams are full, the manager of the dams, presently the *Ministère de l'Equipement et des Transports* (MET) (formerly the Ministry of Public Works), commonly proceeds to release for safety reasons.

The MET is confronted with a dilemma in the management of the dams. On the one hand, they must retain water in order to assure the security of drinking (and industrial) water supply. On the other hand, they must guarantee the safety of municipalities downstream in case of major rainfall. During periods of heavy rainfall, the dam reservoirs reach their maximum capacity and it becomes dangerous to store more water. In fact, the downstream part of the Vesdre basin is regularly inundated due to water released from the dams of Eupen and the Gileppe. People and communes downstream systematically complain to the MET.

The MET found a technical solution to the problem in 1985. They recalculated the mathematical model that helps to manage the filling of the dam reservoirs. They took all the potential users into account. In order to do so, the MET came to the ERPE and asked if they would be ready to face possible shortages in years of exceptional drought. The ERPE, as drinking water producer, was the main user of the water. As they were observing a stabilization in consumption, they consented to the recalculation of the model. As a result, the two actors assumed the risk of water shortages in drought periods, at the benefit of the population downstream who suffered the effects of the regular flood events. The MET reduced the maximum filling level and created a reserve of two million cubic meters in case of major rainfall. The dams no longer threaten the downstream part of the Vesdre. Surprisingly, collaboration resolved the problem in the basin, despite the total absence of legal rules. The arrangement is supported by the consent of the manager of the dams, who has the full capacity to influence the disposition rights to water stored in the reservoir.

The resolution of the rivalry comes with the redistribution of the water resource between the different uses involved. This is due to the consent and attitude of the MET, which evaluated the different risks and uses. The MET plays the role of policy broker between the three (direct and indirect) users involved. At the end of the 1990s, the problem of floods became exacerbated, taking the dams off the agenda. In this case, floods are due to intensive rainfall downstream of the dams. Once again the communes reacted. They used the Vesdre river contract, an informal mechanism of concertation at local level, to set up a working group of local actors that identified the new nature of the floods and their consequences. Furthermore, the kind of informal agreement around the dams was extended to minimum flows for fish and canoeing.

4.3.4 Remaining pollution problems

The pollution of the river Vesdre is certainly the most significant problem in the basin. The part of the river between Verviers and Liège concentrates most of the domestic and industrial discharges. The pollution is historical, with the long-standing industrial activities of the basin (metal, wool and paper industries), nowadays in steep decline. The Vesdre has been a dead river for years. Fish have only reappeared since the end of the 1990s.

In this context, wastewater treatment is a necessity. The main sewers have already been in place since the late 1960s. They eject wastewater directly in the river. Their initial function was to move the pollution away from Verviers and the valley of the Hoëgne. The operator, the *Association intercommunale pour le Démergement et l'Epuration* (AIDE), has acknowledged the problem for years. They planned the construction of the treatment plants but did not have the financial capacity to initiate the construction. In 1977, when they were held responsible for wastewater treatment, no water was treated yet. The regional subsidies on which they depended to perform their task were too small to build treatment plants of a high capacity (above 20,000 population equivalents).

This situation of persistent pollution of surface water provoked a rivalry with an emergent use: tourism. In fact, the pollution prevents the development of tourism, the only economic reconversion expected for this former industrial area. At the same time, urban wastewater treatment has become compulsory.

The resolution of the rivalry came with a convergence of interests between tourist development and wastewater treatment, supported by European subsidies. This part of the Vesdre basin was declared an area of economic reconversion and became eligible for the structural funds of the European Union. These subsidies were then used to develop both tourist infrastructure (e.g. spa in Chaudfontaine) and treatment plants (e.g. Wegnez, downstream Verviers).

Concerning treatment plants, the partial subsidies of the EU have been supplemented by regional subsidies since 1997. Additional funds were raised by the Walloon Region to develop the treatment of domestic effluents under the pressure of the EU. The 1991 EU directive on urban wastewater requires the treatment of all urban effluents as of 1998 (European Communities, 1991). Under this tough constraint, two huge treatment plants finally came into operation and a third one is expected in June 2004.

In this situation, the tourist and the sanitation sector are mutually supportive. The pressure to act comes from the requirements of the EU directive. The flow of funds enables the construction of plants. Then, the improvement of water quality in the basin makes tourist development possible. This dynamic is reinforced with local consultation and co-operation between users: The river contract tries to enhance the natural areas and to emphasise the basin's attraction. As pollution is tackled, tourist activity may effectively become a substitute for industry.

4.3.5 Mineral water production and the environment in the Wayai

The location of a mineral water production plant at the tail of the tributary basin of the Hoëgne-Wayai gave rise to a conflict between the anglers and the industrialists. In its industrial process, Spa Monopole, a private mineral water producer located in the town of Spa, washes bottles and discharges wastewater into the river Wayai. Simultaneously, fish disease occurs. The anglers accuse Spa Monopole of generating the pollution that kills fish. They were demonstrating, while Spa Monopole proclaimed its innocence. A conflict was arising.

In the 1980s Spa Monopole discharged sodium bicarbonate directly into the Wayai. The pollution resulted from the washing of mineral water bottles. Moreover, they caused an accidental pollution with caustic soda. They decided on the construction of a treatment works at the same time. The company is very careful of its brand image. In fact, its activity of mineral water production depends on the protection of fragile aquifers in which it invests much.

In order to restore the water quality of the Wayai and find a solution to the conflict, the anglers' federation (*Fédération des Société de Pêche du Sud et de l'Est de la Belgique)* proposed a river contract with the mineral water producer in 1991. The river contract is a forum of local water users, supported by the communes and the Walloon Region. Users discuss the common problems of the area and take initiatives to tackle them. Actions are determined by consensus and handled by the users on a voluntary basis. The river contract begins with an inventory and stresses the necessary measures to be taken. The inventory could be backed by the biologists of the anglers' federation who were carrying out regular research on the fish population and surface water quality.

Spa Monopole immediately agreed with the proposition. The monitoring network set up by the anglers was reinforced with the assistance of the industrialists. It helped to assess the problem of pressure and more particularly to identify all pollution point sources in the basin. The analyses proved the Spa Monopole was no the cause of fish disease. They showed the innocuousness of bicarbonate emissions and pollution by nitrates was identified as its primary cause. Spa Monopole had nothing to do with that. As a result, the conflict between the anglers and Spa Monopole calmed down and the rivalry was broadened to more water users. In particular, the liability for the pollution was extended to users who did not respect their emission permit, particularly a hotel school and a milk factory. Important leakages from the main sewer were also identified.

The polluters identified committed themselves to take corrective measures in the river contract. For instance, the inter-communal association of sanitation (AIDE) committed itself to repairs of the main sewer. As a result, half of the river contract actions were fulfilled by 2001. The effect on water quality was not significant yet, as the assessment was carried out too early.

In any case, thanks to the river contract, the parties built mutual understanding and common working practises and the conflict was overcome. Co-operation developed and every local water actor got used to considering water as a shared resource. Spa Monopole escaped from a conflict and improved its public image with the Hoëgne-Wayai River Contract. The anglers contributed to improve the river's ecosystem and obtained installations that facilitate fish migration, such as fish ladders. In fact, the real causes of fish disease were identified and gave rise to an organised action from the fishermen and the provision of scientific evidence to back their claim. The resolution of the rivalry was effectively brought by an informal agreement of the two parties on a concerted solution: a river contract limited to the tributary basin.

4.3.6 Successful resolution of rivalries

The Vesdre basin was the location of a series of rivalries between a group of users during the 1980s and 1990s. All these cases occurred in the same area at the same time, but quite independently one from another. The resolution mechanisms have little in common. In Verviers, the enlargement of the uses of water coming from the dam implied a new infrastructure, but no change in former practises. The conformity of drinking water to the public health standard was achieved without modifying the previous industrial use. The solution was implemented only after an offer of guarantees to the industrialists. A reconciliation of the two uses was achieved. Concerning the flood problems, the solution was not brought about by recourse to the law, as in the first case. The owner of the dam arbitrated between the different users in the distribution of the uses. In the third case, the two rival uses were complementary. The use of the Vesdre as a natural environment for tourist activity became possible thanks to the policy of systematic wastewater treatment. The two uses become mutually supportive when it comes to sharing the financial burden. In the case of the river Wayai, Spa Monopole agreed to the fishermen's proposal to collaborate at the scale of the tributary basin. With such an arrangement, the rivalry was enlarged. Spa Monopole proved that it was not the main polluter of the river. Mutual understanding arose from collaboration.

Even if the identified cases of resolved rivalries seem to result from local initiatives, one must not underestimate the EU context and the regional regime that regulates water uses at the local level. Given the difference between the Walloon and the Flemish water regimes, the resolution of rivalries reached in the Dender basin differ greatly from those identified in the Vesdre basin.

4.4 The Dender river basin in Flanders

As in the former part, we present four stories about rivalries between heterogeneous users in a tributary basin.

4.4.1 Presentation of the uses in the Dender basin

The Dender basin is located in the south of Flanders, close to Brussels. The river Dender flows from south to north, with its source in Wallonia. Consequently, the delimitation of the Dender basin is administrative and does not fit with the catchment area. The southern border of the basin is the border between the two Regions. Only half of the catchment area is located in Flanders (708 km^2 of 1384 km^2 in total). The basin that we consider is thus very dependent on the inflow from Wallonia. The main river is the Dender, a navigable river with a rainfall regime that flows into the Scheldt. Its main tributary is the river Mark, which flows along the border with Wallonia. The population of the Dender basin is about 350,000 inhabitants, mainly concentrated between Aalst and Dendermonde.

Figure 4.2 presents a synthesis of the different uses encountered in the basin. We identify urban areas, where the population density is the highest, industries, main agricultural areas (including *polders* and *wateringues*) and nature reserves. We also represent the wastewater treatment plants and the main sewers. Housing is mainly concentrated downstream, as is industry.

Farming activities are mainly localised upstream. Wastewater treatment plants are built along the river Dender and long sewers connect them to the housing areas. Nature reserves are also set along the Dender, particularly between Geraardsbergen and Denderleeuw.

The dominant water uses in the Dender basin are the living environment, production, absorption and protection. Their respective influence must be weighted. Protection, such as flood control, is a rising concern. Water floods are affecting the whole basin as rainfall is intensifying.

The economic structure of the basin is changing. The relative importance of industry and agriculture is in decline as tourism is increasing. Important programs are being developed to enhance the natural environment and the quality of life. First, a major sanitation program is being developed. Second, a policy of land purchase aims at extending the natural areas. Nature reserves are put under the ownership of nature conservation associations (private actors). Third, recreational infrastructure (towpaths, pontoons, marinas, etc.) is being developed. This coupling of infrastructure with water quality should enhance the attractiveness of the basin.

The absence of drinking water production governs the assessment of water management in the basin. In such a situation, the quality aspect is less dominant. Despite a great concern for water quality, no great conflict has emerged. The pressure on wastewater treatment and diffuse pollution (e.g. manure disposal) is less severe. No one intends to produce drinking water from groundwater and the project to produce drinking water from the river Mark has been postponed. Regarding water use, we observe a division of the Dender basin into two parts with the activities that require the best water quality located upstream.

4.4.2 Setting up nature reserves

As in many places in Europe, Flanders is facing a growth in environmental concern. Driven both by the awareness of the population and the requirements of the Wild Birds and Habitats EU directives[4], the Flemish Region has taken voluntary action to create nature reserves on a territory fully occupied by human activities. The solution chosen consisted in developing a network of small protection areas (nature corridors) throughout the Region, taking up land mainly from agriculture. Thus the network of nature reserves is developing in agricultural areas. In the Dender basin in particular, most fields are former wetlands that are continuously drained.

4 Coucil Directive 79/409/EEC of 2 April 1979 on the conservation of wild birds and Council Directive 92/43/EEC of 21 May 1992 on the conservation of natural habitats and of wild fauna and flora.

Figure 4.2*: Identification of rivalries in the Dender basin, 2001*

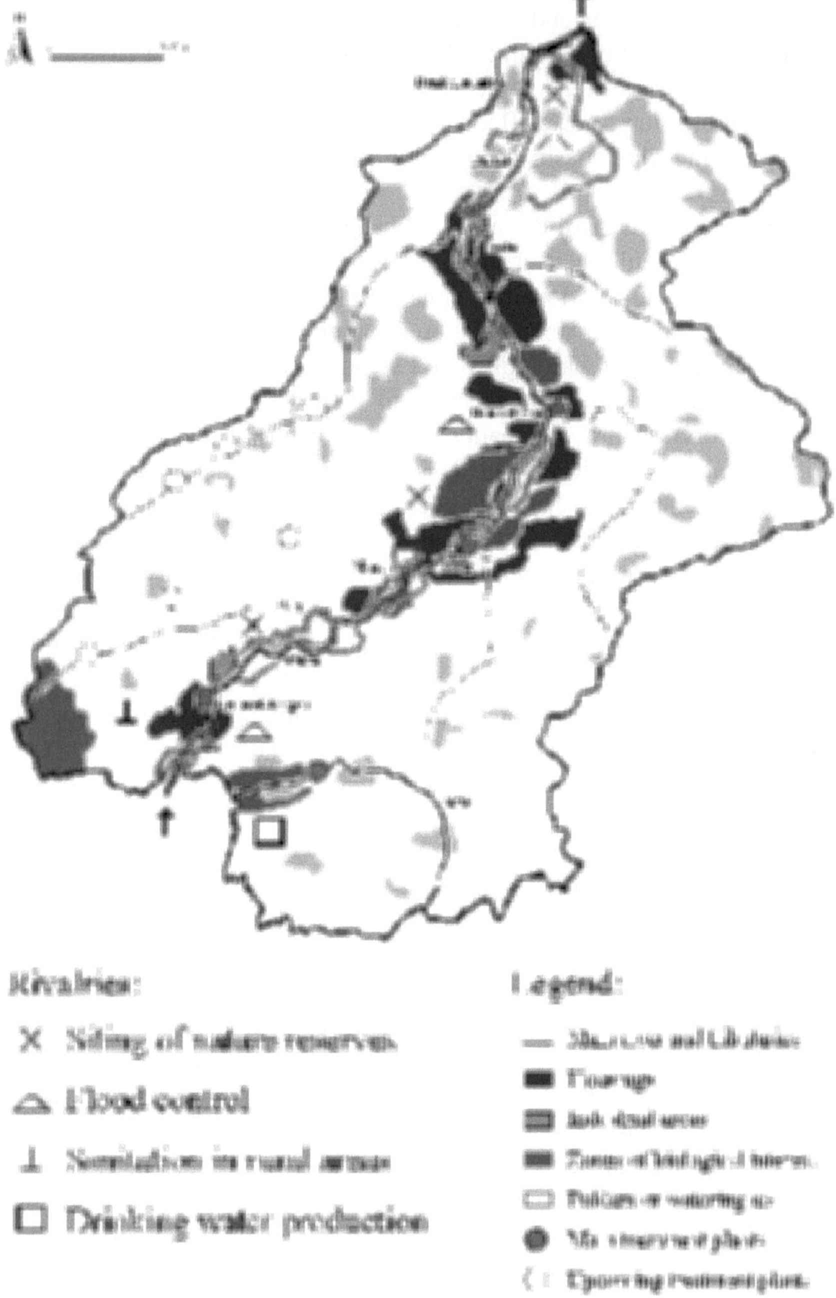

Source: background: VMM and data from AMINAL, VMM and private data collection

Nature reserves are replacing agricultural fields. The situation of rivalry that these nature reserves create between environmental and agricultural uses does not lead to conflicts. The change of land use is accompanied by a change of formal land ownership. Private nature conservation associations are ready to buy land and farmers are ready to sell some fields and to bargain with the associations.

In fact, the changes in land ownership were initiated by the Flemish Region, which subsidises public administrations (AMINAL Afdeling Natuur) and private environmental associations (e.g. Natuurreservaten vzw) in order to acquire fields, with special attention to classified zones of biological interest. Along the Dender, for instance, the nature conservation associations negotiate the purchasing of land, parcel by parcel, directly with the farmers. Expansion of the nature reserves is planned at the regional level, under the supervision of the Region. Priority is given to acquisition in areas of biological interest, classified as such.

Land acquisition is done by volunteers at the local level. Volunteers develop contacts with individual farmers and tend to buy the less productive fields, except in areas of biological interest. In order to ease the transactions, purchasing contracts are sometimes supplemented with more informal arrangements. Once land comes under the protection of the associations, it can still be used as grazing areas. Associations negotiate some purchases in spite of the opportunity to run cattle on their land at no costs.

An absence of confrontation and reasoned purchasing practises contribute to the avoidance of conflict. One use is substituting another. The development of a change in use is made easier with the decline of agriculture. Farming activities are ceasing as the farmers age. Actually, the rivalry is moving into a convergence of interests of the two parties, with public subsidies playing the role of a catalyst. Associations and farmers both benefit from the regional subsidies. The farmers are keen to sell their land as they have received additional resources for their retirement. This targeted concerted mechanism brings money to both parties and, as a result, allows a reorganisation of the territory in favour of nature conservation without aggravating the tension with the farming activities.

4.4.3 Flood protection along the river Dender

All along the river Dender, fields have been drained for centuries to make room for agriculture. The *polders* and *wateringues*, i.e. organisations of farmers, manage drainage over a limited territory. Nowadays, floods are increasing. A combination of significant rainfall and lower porosity of the ground along with expanding housing areas is held responsible for the phenomenon. In order to avoid houses being flooded, the flow of the Dender is diverted onto the fields located on the *polders* and *wateringues*.

The rivalry between drainage and flood protection did not move into a conflict. *Polders* and *wateringues* tacitly accepted that they should drain more. They are there to drain water and they do so. The essence of the organisation is not put into question and the traditional arrangement is strategically maintained. In addition, the Region provides the necessary equipment, i.e. the pumps.

Floods are increasing as rainfall becomes more intense. The flow of the Dender is influenced both by rainfall and by the tides of the Scheldt. During high tides, the river's flow cannot be evacuated from the Scheldt as its level is higher than that of the Dender. The *Administratie Waterwegen and Zeewezen* (AWZ) (the former Ministry of Public Works), i.e. the regional administration that manages navigation and flood protection on the river Dender, is developing solutions to contain the increased flow. As the river is flowing from Wallonia, AWZ cannot manage the volume of incoming water. The problem is particularly severe in the town of Geraardsbergen near the regional border. AWZ has not had the opportunity to build containment areas (a relief basin) upstream of the town. Moreover, the Walloon authorities do not feel concerned about the situation. Thus containment measures are residual and technical. They are limited to the construction of dikes that protect housing areas as water is diverted to the fields, and to equipping *polders* and *wateringues* with pumps.

The fact that water floods are not a permanent problem minimises the rivalry. They often remain limited to the wintertime, a slow period in agriculture. In winter, the cattle are in the barns and the fields are inoccupied. Furthermore, the occurrence of the rivalry is uncertain. The rivalry appears and disappears, is limited in time (not permanent) and is caused by a natural and unpredictable element, i.e. heavy rainfall, for which no-one is held responsible. As a solution, AWZ only brings technical support, i.e. pumps, that eases the drainage of the flooded fields. Actually, the *polders* and *wateringues* never claimed recognition of this competence nor any compensation for the easement affecting the properties of their members. They accommodate with the technical solutions introduced by the Region.

4.4.4 Ineffectiveness of wastewater treatment in rural areas

Surface waters in the Dender basin are severely polluted, as in most parts of Flanders. The process of federalisation delayed the tackling of this regional problem. In 1990, anticipating the pressure to conform to the upcoming EU directive on urban wastewater[5], the Flemish authorities were urged to expand their wastewater treatment infrastructure. This resulted in a systematic policy of investment in water treatment with the single objective of treating 75% of domestic wastewater (29% in 1990). The equipment was standardised. NV Aquafin built high capacity treatment plants and linked the surroundings to them. The size of the sanitation districts is consistently +/-100km². The system had good results in urban areas. Nevertheless, it did not succeed in rural areas. Local people mobilised and protested against the inefficiencies of the investments.

In rural areas, wastewater is too diluted by incoming rainwater and the sewers are too long. In the district of Geraardsbergen in particular, around 100km of pipes are in place. A slow gravity flow leads to self-purification of water with deposits of sludge in the sewers. Water that comes to the treatment plants is already clean. Moreover, during major rainfall the sludge accumulated in the pipes is discharged directly into the rivers through overflows. The pollution is then directly discharged in the river at such moments. As a matter of fact, the sewage system loses its primary utility to improve the quality of surface water as it brings more pollution to the river.

The system developed by Aquafin in rural areas was denounced by a local group of environmentalists, the *Dender Aktiekomittee* (DAK). DAK alerted the public and the regional authorities through spectacular actions and press meetings, complemented by analyses of the water quality in the main sewers. It called for reform of the current sanitation policy. In fact it was not easy to modify the strategy due to the powerful institutional position of Aquafin, the operator of the treatment plants, which enjoys very limited operational liability. Aquafin conducts works that are formally adopted by a tripartite committee comprising representatives of the Region, VMM and Aquafin. Aquafin is only liable to execute works in due time and to respect the emission standards for surface water, whatever the quality of incoming water may be. In Geraardsbergen, the efficiency of the new treatment plant (30,000 p.e.) is below expectation, despite huge investments. VMM and Aquafin, which are behind the investments, reject the responsibility for the poor results. Aquafin states that it only implements investment programs first set by VMM and that the dilution of incoming water is due to a lack of investment by the communes in separate sewage systems. Moreover, the management contract that ties Aquafin to the Region offers it high, guaranteed profits but no obligations of result.

Nowadays, this rivalry has not yet been resolved. The solution brought to the problem of pollution by domestic water remains unsatisfactory in zones of dispersed habitats. Economic imperatives dominate the sanitation policy. On

[5] Council Directive 91/271/EEC of 21 May 1991 concerning urban wastewater treatment.

the one hand, the measures taken are profitable for the water industry[6] and the communes. Communes are discharged from wastewater treatment and take advantage of the sewage system. With regard to Aquafin, the company has made profit on the investments. On the other hand, some people, led by environmentalists, have been pressing for an effective resolution of water pollution. They require a decentralised, integrated water policy and more responsibilities for the communes.

This lack of adequacy between the pressure of the problem and the solutions introduced is explained by a missing link in the institutional arrangements. VMM, who ought to regulate Aquafin's activity, doesn't have the capacity of intermediary between the social demand formulated by the citizens and Aquafin, which brings technical solutions. In fact Aquafin owns the expertise and controls the information. Neither VMM nor AMINAL are in a position to contradict Aquafin. They lack technical expertise to control the private operator. Moreover, the definition of the missions of Aquafin is too vague to allow sanctions in case of inefficiencies. The absence of a resolution to the problem is mostly due to the weakness of the competent authorities.

4.4.5 Production of drinking water on the Mark

Water basin management does not necessarily imply autonomy in water supply. There is no production of drinking water in the Dender basin. Drinking water comes from beyond (Wallonia, Antwerp and Limburg). The project of drinking water treatment plant in the Mark tributary basin was developed, but never went into operation.

In the 1970s, the national water distribution company, the *Société nationale de Distribution d'Eau* (SNDE) developed a project to build a drinking water treatment plant in the Dender basin. The project was an answer to official predictions of consumption increases and serious water shortages (CRPE, 1969). With the federalisation process, it was still carried on as the Flemish Government sought greater independence from Wallonia for the provision of drinking water. The treatment plant was planned downstream of the Mark, the main tributary of the Dender, just before the language border. Expected production capacity reached 20,000 m^3/day and the project included the construction of a water reservoir of 5 Mio m^3.

Despite a favourable political context, the project was abandoned in the late 1980s, mainly because of excessive production costs. It seems not to be the single argument, however. First, drinking water production implies the

[6] One of the main beneficiaries is the construction sector, i.e. companies of public works and design/planning offices. They design and build the infrastructure made in concrete. The corporation of civil engineers conducts the implementation of the sanitation policy.

respect of quality standards for raw water. Although the Mark was classified as surface water intended for the abstraction of drinking water, the required parameters of quality were never met.[7] The completion of the standards required a major decrease of the pollution resulting from domestic wastewater and from fertiliser application. This necessitated tight constraints on manure spreading. Moreover, it was also necessary to convince Wallonia to reduce emissions upstream of the Mark basin, but Wallonia had no interest in doing so. Second, it was necessary to accept a reduction in the area of a protected landscape. The last condition was increasingly difficult to overcome due to the restriction placed on large constructions by environmental permits and the expected opposition of neighbours and environmentalists. Thus, the conformation of the area to the standards would have been dramatically difficult.

Nowadays there is still no drinking water production. Most drinking water supply comes from Wallonia and the remaining part from production plants in Antwerp and Limburg. The northern part of the basin is in fact connected to the mains that supplies the North Sea coast. Thus, an adequate infrastructure of drinking water mains provides drinking water at low costs.

As supply costs are not increasing, this decision concerning the non-production of drinking water is still convenient. It limits the scope of the rivalries that could arise within the basin, in particular the problem of the pollution of surface water by the nutrients spread by farming activities. The liability to preserve potential drinking water is at some point transferred to external areas. As a matter of fact, the enhanced requirements of the water framework directive in terms of drinking water production are 'externalised'. The rivalry is not resolved, but displaced out of the basin.

In fact, the decision to abandon the construction of the drinking water treatment plant in the Mark valley was the result of arbitration between economic and political stakeholders. Initial conditions were bad too: the project was not profitable; it was difficult to regulate manure disposal; and water production depended on the quality of inflows from Wallonia. As a result, the policy of hydrous independence displaced its effects. The capacities of drinking water production were preferably developed in the North and completed with mains connecting the other parts of the Region.

[7] The parameters are set in respect for the Council directive 75/440/EEC of 16 June 1975 concerning the quality required of surface water intended for the abstraction of drinking water in the Member States. Surprisingly, the river Mark is still classified as such and the new wastewater treatment works built in the area (e.g. Galmaarden) respect the specific standards.

4.4.6 Mitigated success in the resolution of rivalries

The analysis of water management at the scale of tributary basins allows us to consider particular cases of rivalries between heterogeneous users. We have related four stories that attest to divergences in the mechanisms by which the rivalries were resolved. In the first case, the expansion of natural areas was made possible with a substitution of a rising use (nature protection) for another in decline (agriculture). Financial support for the substitution and a joint opportunity of the two categories of users avoided conflicting situations. In the second case, the two uses continue without much rivalry. The groups of users affected by floods accommodate to the disturbance despite a worsening of the situation (intense rainfall). In fact, the rivalry is occasional and occurs mainly in winter when farming activities slow down. In the third case, the problem was not resolved, despite the implementation of a voluntary policy supported by huge investments. The problem, however, is limited in space, i.e. to rural areas, and the Flemish policy against pollution of surface water gave globally satisfactory results. In the last case, the rivalry was moved out of the basin. Drinking water production was displaced to other areas and thus rivalries concerning the quality of surface water did not arise inside the basin. Each case of rivalry is approached in a particular way, involving specific, targeted measures.

4.5 Comparative analysis

The eight stories told reveal the importance of considering the EU context, the regional water regimes and the way rival users activate the institutional rules of these regimes. Users are conscious of the existence of both property rights and public policies affecting water uses and protecting their share in the distribution of uses. We now consider more systematically the way the rules of the regimes influence more sustainable water uses. We conduct two kinds of complementary comparisons. The one identifies the regularities in the resolution of the rivalries we have presented, independently of considerations of the basin in which the rivalries are located. The other focuses on the characteristics of the basins and regional regimes in force into each of them. We look at the way water is managed and set the conditions for a more sustainable water management in each case. These two comparisons allow us to stress preliminary conclusions on the recourse to property rights and public policies by the users involved. Finally, we put the challenge of enforcement of the WFD at the tributary basin scale according to the institutional logic that prevails in the basin.

4.5.1 Conditions for the resolution of rivalries

The first comparison isolates the successive stories about resolutions of rivalries and compares them without any consideration to the specific natural and regulatory environment of the basins. We start from a logic of local use (bottom-up approach). The stories told are rivalries that involve different kinds of water users. The eight rivalries investigated also emerge from very different problems, from the pollution of drinking water to floods or the constitution of nature reserves. Nevertheless, in each case actors seek solutions, even through conflict, and mobilise particular elements of the regimes, be they property rights or public policies.

Table 4.1 presents the respective influence of property rights and public policies in the resolution of the local rivalries. The main thing we learn is that the resolutions of rivalries emerge either from recourse to policy design or recourse to property rights by the local water users. Some rivalries are resolved by recourse to the policy design. In the story of lead poisoning in Verviers, local citizens mobilise the quality standards of drinking water, placing them before the Court. In the story of tourism in the Vesdre valley, two uses -- tourism and purification -- mutually benefit from prescriptive rules on domestic wastewater treatment and incentive measures for economic development. Some other stories are resolved by recourse to property rights. In the case of water floods in the lower Vesdre, the rivalry occurs between the needs of both industry and the drinking water producer and the damage to the victims of water release from the dams in the lower valley. The manager of the dams, i.e. the Ministry of Public Works, is entitled to formal ownership of the dams. In the absence of any regulation on water release and minimum flows the manager of the dam plays the role of policy broker between the users involved and redistributes disposition and usage rights by mutual consent. In a different context, the siting of nature reserves -- a rivalry between agriculture and nature reserves -- is resolved by a modification of the property rights structure on land. In this case, a public policy is conceivable, but the legislator privileged a change in property rights. This option gave satisfactory results. As a matter of fact, none of the conditions, i.e. a recourse either to policy design or property rights, is a necessary condition for the resolutions of rivalries. We also deduce that the combination of recourse to policy design and property rights is not necessary for conflict resolution.

In six cases out of eight, we observe a resolution of rivalry with recourse either to policy design or property rights. In two cases however, the rivalry is not solved. In one case the problem is displaced. The story of the abandoned project of a drinking water treatment plant does not mobilise any element of

Table 4.1: *Elements of the regime influencing the resolution of rivalries*

Rivalries							
Resolution						No-resolution	
through Policy Design			through Property Rights			Unsuitable solution	Moving of the problem
Lead-poisoning (Vesdre)	Tourism in the Vesdre valley	Fish disease (Vesdre)	Flood control along the Vesdre	Siting of nature reser-ves (Dender)	Flood control along the Dender	Sanitation in rural areas (Dender)	Drinking water (Dender)
• European and Belgian stand-ards on drink-ing water • **Action in jus-tice** from a group of citiz-ens against the provider • Case-law of the European Court of Jus-tice • The Central State (the Region since 1980) finances and builds the drinking water treatment plant	• **Joint chance** between tourism and sanitation. Financial flows for both uses • **Obligation to treat urban wastewater** • European Regional Funds for tourist development • Local concert-ation between actors in the **Vesdre River Contract**	• Local co-oper-ation in the Hoëgne-Wayai River Contract • Extension of the initial conflict to the whole basin and users • Development of dialogue and mutual understanding between the parties	• The **owner of the dams, i.e. the State**, distributes usage rights linked to the dam reser-voir (drinking water, energy, strategic reserve and protection) • Voluntary action of the owner of the dam that plays the policy-broker	• Change in land ownership transforms former fields in nature reserves • **Purchase by environmenta l associations**, subsidised by the Region • Local negotia-tion between nature associa-tions and ind-ividual farmers • The Region determines the priority zones and gives its agreement to purchase	• **Easements** on the territory of the polder. The polder pumps flooded fields • **Dike management** by the Region with full ownership. Capacity to expropriate • Financing of the pumping activities of the *polders* and *wateringues*	• **Regional sanitation policy** • Standardised infrastructure with low effectiveness in rural zones • High pressure from an environmental association to modify the rationale	• Classification of the river Mark as potential drinking water • Flemish policy of hydrous independence • 'Externalisa-tion' of drink-ing water provision and reorientation of investment in other areas

the regime. In this case the rivalry is avoided with the freezing of the project. The rivalry does not emerge, as the demand for a redistribution of uses is not even formulated. In the other case, the solution brought by the policy design does not fit with the local environment. In Geraardsbergen, the implementation of the Flemish sanitation policy does not lead to satisfactory results. Wastewater from the surrounding of Geraardsbergen is still not effectively purified, despite a huge investment in wastewater treatment. Implementation of regional plans at the local level is ineffective if it does not fit with the local characteristics and environment. Finally, the recourse to policy design does not necessarily resolve the rivalry, so it seems necessary to systematically observe both elements of the institutional resource regime. If one only considers one aspect, one is immediately faced with a series of cases that cannot be explained.

The choice that rival water users make between recourse to policy design or property rights remains to be understood. According to our own interpretation, we could consider that a mobilisation of the property rights is made mostly in quantity cases. There the property rights make the link between land and water, notably through the right of accession. The matters of allocation of water, quantitative aspects and redistribution may be more often considered through property rights. In our stories, changes in property rights resolve rivalries concerning water floods and wetland areas. They are quantitative problems. By contrast, quality problems are resolved with a mobilisation of the policy design, e.g. sanitation and quality of drinking water. Rivalries around water quality may more often mobilise the public laws, such as the European directives and regional laws. If these hypotheses were to be verified, it would be a bad thing if the European Water Framework Directive (WFD) does not take the property rights aspects into account as it reforms water quantity management (see Aubin and Varone, 2004).

It appears that neither a modification of the policy design nor a modification of property rights are necessary or sufficient conditions in the resolution of a rivalry between heterogeneous water users. From each situation peculiarities emerge in the answer to rivalries, either intervention with the policy design or with changes in property rights. This is evidence of the pertinence of the institutional resource regime in order to analyse water management mechanisms.

The resolution of rivalries, taken as isolated stories, in many cases comes from the mobilisation of elements of the institutional regime by the users involved. This is the case with Verviers, where the citizens activated the law with Courts action. The manager of the dams in the Vesdre and the environmental associations in the Dender mobilised the property rights system to push through local adaptations. In some other cases, arrangements were put in place by the responsible authority without much concertation. The respective weight of changes in policy design or property rights cannot be balanced with the present results.

4.5.2 Conditions for sustainable water use

After an analysis of the broad triggers and conditions for the resolution of particular local rivalries, we observe in a broader way the water management in the two tributary basins. Following the institutional definition of the basin for WFD implementation we wonder if the basins are managed in such a way that the water uses are sustainable or not. The promotion of sustainable water use is a purpose of the WFD. Sustainability is achieved when the variety of heterogeneous water uses are satisfied in a socially acceptable way and without threatening the capacity of the whole resource to satisfy these uses in the long run. The satisfaction of heterogeneous use requirements must preserve the resource stock (e.g. wetlands) and qualitative improvement of water (e.g. purification).

If we look at rivalries in the two basins, we observe that resolution occurs more often in the Vesdre basin than in the Dender basin. Systematically, the public policies and property rights both play an important role in the two basins. In the Vesdre basin, despite initial conflicts between actors, the capacity to reach local arrangements partly determines the resolution. In the Dender basin, changes, and consequently conflict resolution, come from the top, i.e. the regional administration. In most cases the rivalries are not solved sustainably. There the characteristics are no longer linked to the rivalries, but to the institutional logic.

The main differences in characteristics between the two basins are the degree of dependence from outside and the production of drinking water. The Dender basin is very dependent on Wallonia. Inflows are important as the southern border of the basin does not correspond to the watershed, but to the administrative border between the two Regions. The dependence of the Dender basin is reinforced by the necessary imports of drinking water. The absence of a production plant in the basin consequently diminishes the pressure on the overall water quality. Moreover, groundwater is not subject to particular protection measures. The situation is totally different in the Vesdre basin where there is significant production of drinking and even mineral water. Thus, the characteristics and problem pressure differ widely between the two basins.

Natural characteristics and objective problem pressure are not sufficient elements to explain the difference in water management between the two basins. The basins are subject to diverging water regimes (cf. section 4.2). In

Table 4.2*: Comparison of the characteristics, problem pressure and regimes between the two tributary basins of the Vesdre and Dender*

	Vesdre basin (Wallonia)	Dender basin (Flanders)
Characteristics of the basin	Independent basin Administrative borders fit the 'natural' watershed Rainfall regime Local identification to the basin	Hydrous dependence Administrative borders do not fit the watershed Rainfall regime No cultural identification with the basin
Uses and rivalries	Production of drinking water Low water quality (industry + domestic use) Strong need for wastewater treatment Water floods Development of tourism Fishing	No production of drinking water Poor water quality (agriculture + domestic use) Strong need for wastewater treatment Water floods Nature reserves
Regional regime	Subunit of management WFD Great autonomy of the local level High degree and visibility of social conflicts Lack of planning	Subunit of management WFD Centralisation at the regional level Low degree of social conflict Planning: policy choices formulated at the regional level
Mechanisms of resolution of rivalries	Legal action Informal arrangements between local actors Joint implementation of sanitation policy and development of tourism under EU incentives	Changes of land ownership, with subsidies Management of dikes and acceptance of easements by the farmers Sanitation policy with major regional investment Freezing of the drinking water treatment plant project before the rivalries emerge
Challenges of the WFD	→ *Management and monitoring programs* → *Basin authority*	→ *Negotiated solutions* → *Local participation* → *Basin identity*

the Vesdre basin, the local level is very autonomous. For instance, the construction of wastewater treatment plants is decided at the scale of the Province by inter-communal associations. This relative autonomy in the Vesdre basin leads to more informal arrangements between local actors. The communes play a greater role in water management. For instance, they lead the inter-communal water associations and the local mechanisms of collaboration, i.e. the river contracts. Concerning the Dender basin, decision-making procedures are more centralised. Water management is more subject to regional planning. Local mechanisms of collaboration, i.e. the basin committees, are less developed and put under the supervision of the Region.

As can be seen in Table 4.2, the water framework directive (WFD) ought to be applied in Belgium at the level of the tributary basin. At least functionally, it should be the unit of enforcement both in Wallonia and Flanders. Both the Vesdre basin and the Dender basin, then, are subunits of management for the WFD. However, the two basins present wide divergences both in terms of initial characteristics and the regimes developed. The Vesdre basin is managed in a bottom-up manner, while the Dender basin is managed from the top, i.e. from the Region. The first is more often crossed by conflicts of use than the second. Thus the differences in structures (morphology, institutional arrangements resulting from the regional regimes) are fundamental to an understanding of the resolution of rivalries towards more sustainable water uses.

In terms of sustainability, we admit, that the regime in the Dender is not 'better' than the regime in the Vesdre. Over the studied period, the water quality of the rivers was improving, roughly speaking, even if the pollution remains acute in both basins. At first sight, we could imagine that more planning and more intervention in the property rights system would automatically introduce a more integrated and consequently a more sustainable water management. In the Dender basin, centralised regional planning brings standardised answers to local problems, an arrangement that lacks acceptability, efficacy and efficiency in particular cases. Moreover, the concentration of responsibilities in the hand of the regional administration does not automatically lead to co-ordination and integration.

The two basins appear to be extremely different, and in both cases the WFD ought to bring improvements. This presupposes an adaptation of the required measures to the characteristics of the tributary basins, including the local interests and informal rules. In the Vesdre basin, the local actors activate institutional rules of higher levels in order to resolve their local conflicts. In several occurrences, the Walloon Region, the European Commission, the Structural Funds and even the European Court of Justice have been mobilised. Local actors do not hesitate to go to court in order to solve their dispute. The recourse to authorities at a higher level attests to a lack of vision. The basin lacks a top-down approach that could plan action and corrective measures. The challenge with the WFD implementation will consist in using instruments of planning above the local (and often informal) arrangements. The main issue at stake is the development of a management plan. According to the WFD, the management plan consists in aggregating in an orientation document the characteristics of the basin and the pressure exerted by water uses. This would compile monitoring data and indicate, at the scale of the basin, which measures are necessary to guarantee good water quality. The management plan is being made by a centralised authority that gets a global overview of the situation and problems in the basin. The basin authority can thus coordinate action to achieve better results. Nowadays, water management is mainly local, with mutual agreements. It consists in negotiations about point-source problems that come to (informal) agreements. In such a context, how could an efficient and acknowledged management authority emerge? Planning must be improved.

The situation is rather different in the Dender basin. Planning is already well developed. However, this does not seem to be sufficient to get integration. The introduction of bottom-up participation is crucial. A greater involvement and participation of the local actors seems necessary. Water management is initiated at the regional level, but once put into practice, local actors need finer mechanisms to resolve their disputes. Evidence comes from the stories presented above. The *Denderaktiekomittee* protested against the sanitation policy managed by Aquafin. The committee went to the Flemish Parliament to be heard on the inefficiency of this policy in rural areas. Farmers in the polders discuss with officers of AWZ, volunteers of the environmental associations conclude contracts about cattle grazing with farmers as auxiliary to land purchase. In the context of the WFD implementation, the central stake in the Dender basin, then, is to introduce more local participation in a tributary basin that lacks identity. The Dender basin has borders, but they do not fit with the watersheds. There is no identification of the population with the basin.

The progressive introduction of the principle of full-cost pricing is one of the pillars of the WFD. Full-cost pricing arises from the idea that every direct or indirect water user should pay the estimated price of the pollution that he causes to the water resource. In our cases of rivalries no single resolution results from the introduction of price mechanisms. If the price of water is maintained or increased, what then is going to be modified?

Finally, the implementation of the WFD should take account of the existing property rights and public policies on the water resource. Each basin has its own characteristics and regime. Thus, implementation measures should be appropriate to different settings. Integration, and more particularly sustainability, cannot be achieved by uniform measures. In each case, the conditions of a sustainable water resource management are rather different.

4.6 Conclusion

Water management in Belgium is highly differentiated from one local place to another. The difference is first and foremost due to divergence in the development of the regional water regimes. The present analysis of two tributary river basins, the Vesdre in Wallonia and the Dender in Flanders, gives a detailed perception of the questions at stake in the development of integrated water management. In the present chapter, we have identified the conditions that contribute to a resolution of rivalries between heterogeneous water users and tried to see, inside a given basin, if the resolutions contribute to a more sustainable management of water. Local users distinctively mobilise the policy design and property rights to solve their dispute. However, we cannot state the conditions under which the mobilisation of one element has primacy over the other. Mobilisation of one or both elements is a necessary condition for success in the resolution of rivalries, but not a sufficient one, although successes in the resolution of rivalries seem to bring more sustainability to water uses. The quality of the water has improved over the period. The best illustration is the reappearance of fish in the Vesdre and Dender rivers in recent years.

Finally, each case of rivalry has its own peculiarities, and rivalries cannot be overcome with one general recipe. Conditions must be carefully considered for every specific case. They are tied to the characteristics of the basin and the institutional water regime in place inside the basin considered. The regime is mobilised by the local actors to find ways to resolve their rivalries.

The WFD directive will certainly not solve all potential rivalries connected to water. Its general goal aims at integrating water management and reaching sustainability and wise use of the water resource. Up to this point, the implementation phase of the WFD should avoid exacerbating conflicts between users and cause more fragmentation to the current regimes through authoritative arrangements. The general conditions in which the WFD will be implemented are decisive, especially because the property-rights system in place is not identical from one basin to another.

Management at the level of river districts is questionable if it does not take the lower levels into account, i.e. the tributary river basins. From our analysis of local water management in Belgium we observe that many problems are specific to a Region or even to a tributary basin. The consideration of the whole district for water management is necessary but not sufficient. When only considering water at the district level, potential local rivalries are missed and specific, huge problems are left unsolved or even aggravated. A centralised planning, ignoring the existing regulative systems and local conditions, misses effectiveness when transposed to local places. If not complemented by effective structures of local participation, the promised water management at the level of the river district cannot meet good status criteria for the quality of surface water nor further sustainable water uses.

Most of our findings corroborate the broader comparison made in Chapter 9 between several European catchments. The authors state, from experience of a more integrated water management, that some variation of the institutional arrangements must be accepted. To a certain degree, differentiated local settings may contribute to the good status of water with more effectiveness than standardised answers.

REFERENCES

Aubin, D. and F. Varone (2004) The Evolution of European Water Policy. Towards Integrated Water Resource Management at EU Level, in: I. Kissling-Näf and S. Kuks (eds.) *The Evolution of National Water Regimes in Europe. Transitions in Water Rights and Water policies towards Sustainability,* Dordrecht, Kluwer Academic Publishers.

Aquafin (2001) *Annual Report and Environmental Report 2000*, Aartselaar, Aquafin, 70 pp.

Association des communes du bassin de la Vesdre (1999) *Projet de contrat de rivière Vesdre: dossier préparatoire*, 52 pp. + annexes.

Commissariat royal aux problèmes de l'eau (CRPE) (1969) *Le problème de l'eau*. Rapport final du Commissaire royal (3 vol.), Bruxelles, Service du Premier Ministre, 250 pp. (unpublished).

Contrat de rivière Hoëgne-Wayai (2001) *Evaluation finale* (unpublished).

European Communities (1980) *Council Directive 80/778/EEC* of 15 July 1980 Relating to the Quality of Water Intended for Human Consumption.

European Communities (1991) *Council Directive 91/271/EEC* of 21 May 1991 Concerning Urban Waste Water Treatment.

Ministère de la Région wallonne (1995) *Atlas de l'eau de la Wallonie,* Namur, DGRNE, 50 pp.

Ministerie van de Vlaamse Gemeenschap (2000) *Omgevingsanalyse Dender*. Project Functieplannen Startrapportering, Aalst, AWZ Afdeling Bovenschelde, 87 pp + annexes.

Ministerie van de Vlaamse Gemeenschap (2000) *Op weg naar integraal waterbeheer. Het watersysteem in het bekken van de Dender*, Brussel, AMINAL, 64 pp.

Vlaamse Milieumaatschappij (2000) *Algemeen waterkwaliteitsplan Dender*. Samenvatting, Erembodegem: Vlaamse Milieumaatschappij, 64 pp.

places. If not complemented by effective structures of local participation, the primary water management at the level of the river district cannot meet good status criteria, for the quality of surface water nor for the quantitative water uses.

Many of the findings [illegible]

REFERENCES

Chapter 5

An Innovative but Uncompleted Integration Process

The Audomarois and the Sèvre Nantaise in France[1]

Jean-Marc Dziedzicki and Corinne Larrue
Université François-Rabelais (Tours-France)

5.1 Introduction

France is a country that is well provided with water, with little dependence on water from neighbouring countries. A multitude of watercourses flow through the country, including about 30 major rivers are more than 200 km long, to which must be added the Rhine that for 190 km serves as a frontier in the East. Moreover, running surface water includes several thousand kilometres of artificial canals of which 4,500 km are classified as navigable. The importance of this network of waterways underlay the formation of six agencies created by the law of 16 December 1964, which play a major role in the management of water in the main drainage basins: Adour-Garonne, Artois-Picardie, Loire-Brittany, Rhine-Meuse, Rhône-Mediterranean-Corsica and Seine-Normandy. This overall situation takes no account of the differences in the distribution of water resources between regions, nor the general problem of water quality, which is often poor. The regions with most rainfall have four times as much as these with the least, while the four major rivers (the Seine, the Loire, the Rhône and the Garonne) account for 60% of surface water resources. The general balance between resource and needs is therefore precarious in some places, the situation in some areas being critical. Indeed, the intensification of certain uses has led to a deterioration of water quality, due particularly to industrial, household or agricultural

1 This text is based on the case study analysis conducted by I. Calvo and I. Verdage, and on a legal framework analysis conducted by I. Sangaré.

Hans Bressers and Stefan Kuks (eds.), Integrated governance and water basin management. Conditions for regime change towards sustainability, 131-164.

pollution. At the same time, in addition to the demands of traditional uses, the development of new uses, notably those linked to leisure activities, requires quantitative and qualitative water management systems to be set up, which will gradually transform the way water resources are perceived and organized.

Current water management in France is thus characterized by specific regional situations: in some regions the balance between resource and needs is critical (mostly in the south west), in others the main problem is linked to an overuse of surface water (main estuaries) or with diffuse pollution stemming from agricultural activities (Brittany, Parisian basin). Most rivers have been 'artificialized': large ones (for hydro-electricity or cooling purposes) as well as small ones (for irrigation or drainage purposes). But thanks to a 1992 law that promotes the principle of balanced management of the water resource, and to the growing attention given to environmental issues, the last decade has seen a gradual change in the organisation of river and water management. The European Framework directive will reinforce this trend.

The aim of this chapter is to explain the integrative approach that has been developed since the 1964 Law in France (section 2) and the way this national framework has helped in managing conflicting water uses at the river basin level with the examples of the Audomarois and Sèvre Nantaise river basins (section 3 and 4). We analyse lastly the difficulties and the promises of implementing integration of property rights system and water protection public policies (section 5).

5.2 The French national water regime

The development of the water regime in France is based around two main elements: the definition of property rights, the structure of which was laid down in the law of 8 April 1898, and the setting up of public policies to control and manage the multiplication of uses, from the 1960s.

Concerning the *definition of property rights*, the law of 8 April 1898 remained, even after many reforms, the basic text concerning the legal regime of water. It would dictate all developments in water law, centred round a single idea: to reduce ownership without calling it into question. Under this law:

1. Ownership was recognized for rainwater, spring water, ponds and canals, with a particular leaning towards right of disposal. Moreover, in the case

of spring water, this was subject to numerous exceptions which seriously restricted the general principle of ownership;
2. Uses of navigable and floatable rivers (under public ownership) were subject to compulsory authorisation;
3. For non-navigable waterways, the law introduced a crucial and original innovation: the dissociation of the bed from the water flowing over it. From thenceforward the bed belongs to the riparian owner, but the water on it belongs to nobody and can only be subject to user rights, according to article 644 of the Civil Code. The consequences of establishing the riparian resident as effective owner led to the gradual increase of administrative rules and checks in order to limit the scope of the principle arising from this regime of non-navigable watercourses. In this way, water rights became increasingly complex and intractable, with a multitude of texts, which were supposed to meet a variety of situations covering both the water regime and uses.

The legislature sought to encapsulate the strict principle of ownership of water, on the one hand by imposing various easements to facilitate certain uses or respect certain aspects of the natural water cycle; and on the other hand, by developing administrative law to the detriment of private right in matters of water management. For example, there was a rapid increase in systems of authorisation, such as the one introduced by the 1919 law on hydro-electricity. The law of 3 January 1992 is remarkable for generalising a single system for declaring and authorising all non-domestic withdrawals and discharges, and for all uses that could have an influence on the water system.

Concerning *public policies on water management*, their development can be summarised in terms of three main periods circumscribed by major legislation: from the 1898 Act up to the middle of the 20th century, a series of particular regulations (navigation, hydroelectricity, public health, etc.) led to what can be considered sectoral management of water. After World War II, the strengthening of State intervention and an awareness of water as an important issue in managing economic development led to the initiation of a system of global management of water resources (distribution and anti-pollution measures) with the Water Act of 16 December 1964. A notable innovation of this Act was to set up water agencies and basin committees. The gradual emergence of new concerns since the 1970s (the creation of the Ministry of the Environment, the Nature Protection Act of 1976, the Fishing Act of 1984) has given a higher profile to the protection of the natural environment in matters of water management. Lastly, the 1992 Water Act promoted the idea of balanced management (integrated management), based notably of the creation of new planning and negotiation tools (SDAGE and SAGE[2]), and new institutions at local level: the Local Water Commission composed of local authorities, state public administrative bodies and user representatives. This new law brought to light problems concerning the integrated management of water at basin level.

The water regime evolution in France can thus be summarized as follows:

- 1566-1789: simple regime, with few uses (navigation, water supply), and division between royal right on major rivers and feudal right on other surface water;
- 1789-1898: simple regime, with few uses, but the appearance of some new uses (irrigation, industry, water treatment and waste disposal) and disputes linked to the absence of clear laws concerning non-navigable or floatable water, the status of which was clarified in the law of 8 April 1898;
- 1898-1945: diversified and complex regime (sectorisation), with management of different uses weakly coordinated (supply, agriculture, navigation, hydro-electricity, etc.);
- 1945-1964: complex regime with increase and particularly intensification of uses linked to industrialisation and urbanisation, but the appearance of the first elements to coordinate different uses (new actors such as EDF and regional development companies);
- 1964-1992: 'global' regime, weakly integrated, ensuring coordination between different uses, notably at the drainage basin and national levels; preservation of quality by fighting pollution, and limitation of conflicts over ground water between industrial uses and water supply; first steps to 'polluter-payer' and 'abstractor-payer' system; assessments (regular pollution checks of watercourses) and experimentation (river contracts from 1981); consideration of ecological quality of the water environment (cf. 1984 'fishing' law), etc.
- since 1992: 'balanced' regime, directed towards local integration centred on the resource and its quality: fight against diffuse pollution (particularly agricultural); search for better application of 'polluter-payer' principle; application of scheduling: SDAGE (Master Water Management Plan, obligatory for the six drainage basins) and SAGE (Local Water management Plan, optional and depending on the will of local authorities round the sub-basins). Intervention methods in water management thus evolved under the influence of various factors: level and scale of organisation; a richer network of actors; co-ordination of the

[2] SDAGE: Master plans for water development and management (*Schéma Directeur d'Aménagement et de Gestion des Eaux*); SAGE: Local plan for water development and management (*Schéma d'Aménagement et de Gestion des Eaux*)

multiple viewpoints affecting water management; use of different means for developing the policy; making those involved take responsibility for reaching objectives.

Nevertheless, the Water Act of 1992, as that of 1964, did not touch on the legal framework of water ownership. Debates and preparatory work on the law had nonetheless envisaged a reworking of the general framework. The Water Conference had put forward two bold solutions but these were not generally accepted. In reality the Act of 3 January 1992 did not go as far as had been expected during the preparatory meetings. However, by proclaiming water to be an object of national heritage, the legislator recognised, for the first time, the need to protect this resource for itself, as much its quantity as its quality. The codification of the requirements of this new concept could lead to new legislation, which would define more precise measures for an efficient management of water heritage.

Finally, as far as institutions are concerned, the main levels of organisation of water management can be classified as coming under the responsibility of:

1. the European Community, which address directives to which national legislation must conform;
2. the State, which draws up rules and general principles of management (laws, decrees, by-laws etc.) and ensures administrative coordination of state actions and actors (by means of the creation of a new level of decision within State representatives (at the water basin level);
3. the six Water Agencies and the six basin committees (consisting of representatives of the State, regional bodies and users for each basin) which were created by the 1964 act (and transformed by the 1992 one) which ensure resource planning and management, as well as financial solidarity through pollution and consumption fees;
4. Communes and inter-communal bodies (i.e. at the level of a sub-basin, amalgamating several communes) that ensure service provision, and technical, administrative and political management. The level and scale of local management are therefore very varied, as are the objectives, methods and instruments. That of the actors intervening in water management matches this diversity. Some of these actors, such as mayors, have had increased responsibility since the 1992 Act.

The following sections describe two case studies of river basin management in France, in order to present local contexts of implementation of French national water regime.

5.3 The Audomarois basin

5.3.1 The Audomarois drainage basin and the current situation

The Audomarois is located in the North of France. This region has a history of traditional and heavy industry and has experienced a profound economic crisis in the last forty years (See Figure 5.1). Audomarois is an area of approximately 620 km² that corresponds to the whole of the drainage basin of the River Aa, i.e. the catchments area of the river and its tributaries, as well as its floodplain -- the Audomarois marsh (near the Saint-Omer *commune*). Part of this area is included within the boundaries of a Natural Regional Park (the 'Park'), which was officially created in 1986. In general, the Audomarois is a rural area but with a fairly well defined urban zone in the East. The area includes 72 *communes* with a total population of 97,231 inhabitants (Insee 1999).

Figure 5.1*: The position of the Andomarois Basin in France*

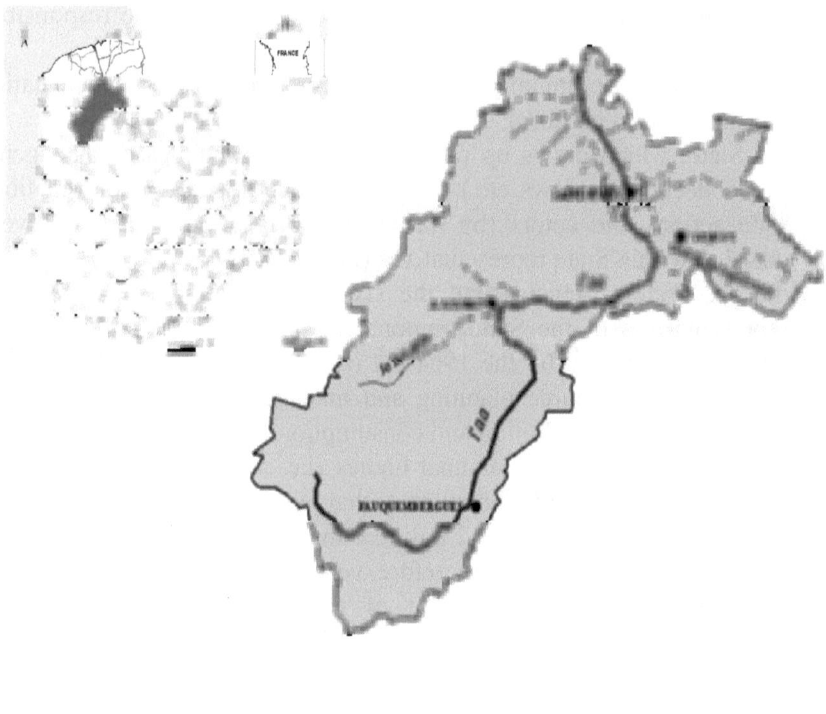

The Audomarois is in fact a fairly complex area from a hydrological point of view, with three specific water bodies:

- *Groundwater* is present throughout the Audomarois to the extent that this area is considered as the 'water tower' of the whole region. The River Aa and the marsh are the main natural outlets.
- The *River Aa* is about 50 km long. Along the Aa and its tributaries there are 46 small dams, the main purpose of which is for fish farming upstream. Downstream, most of them have been abandoned. The Aa is not navigable and therefore belongs to the private sector, coming under the regime of privately owned watercourses.
- The *Neufossé canal* (public domain) crosses the marsh and can be seen as the regulator of the water level in the marsh since it was built in 1756. With the building of the canal, the course of the Aa up to its mouth was canalised, thus modifying the relationship between the river and the marsh. A network of small canals called *watergangs* (water lanes) in fact links the canal and the marsh to each other. This canal is the only outlet of the Audomarois basin to the sea. The Neufossé canal comes under the regime of (public) state-owned watercourses, which gives ownership of the bed and management of user-rights to the State.

These three water bodies are connected, but there are still some uncertainties about their connections. The main connection is the *Audomarois marsh*. This is a wetland, which is the heart of the basin. The 3,400 km^2 of the marsh are drained by 160 km of *watergangs* that depend on the private sector (regime of privately-owned watercourses). This marsh was farmed and therefore managed from the 7th century up to the beginning of the 19th century. Today it is an entirely artificial wetland, under human control.

While attempts to drain the Audomarois marsh were made as early as the 7th century, the first planning works really started in the 12th century, undertaken by monks and land-owners (drainage, routing of new canals, dikes, etc.). In the 18th and 19th centuries, they further made their mark on this area by carrying out works inspired by Dutch techniques. As the market gardening sector developed in the 19th century, small industries were set up, helped by a favourable hydrological situation, which enabled them to make use of good quality ground-water. But, especially during the 1960s, this area underwent important changes, leading to an unstable situation in the seventies: agriculture and market gardening became specialised, modernised and intensive, the growing discharge of polluted water from industrial activities had a negative impact on the natural environment, a rising demand for drinking water in the Dunkerque district led to the drilling of more boreholes into the water table, water-related leisure activities developed unchecked, Saint-Omer (the largest city in the area, situated on the edge of the marsh) continued to grow, spreading onto the marsh, etc. Various activities in the area were no longer sustainable, and conflicts related to water use emerged.

Today, the management of these conflicts is set up through the formulation of a Local plan for water development and management in the water basin of the River Aa (Schéma d'Aménagement et de Gestion des Eaux -SAGE-). The SAGE formulation process has been set up since July 1992 on behalf of the Park (six months after the publication of the new water law that created the SAGE tool). The objectives given by the Park to be addressed by the SAGE are genuine improvement of water quality for the River Aa and its drainage basin and conservation of the drinking water supply. A groundwater study was carried out in 1994 and its results were published in 1995. But the SAGE process really began in 1995[3] with the creation by the Préfet[4] of the Local Water Commission (Commission Locale de l'Eau -CLE- composed of 20 members from local authorities and local public establishments, 10 members of State public administrative bodies and 10 user representatives). The role of the CLE is to draw up the SAGE. SAGE then entered the preparatory phase, at the end of which the 'SAGE document' should have been produced through the consultation process and approved by the Préfet. The consultation process started through the CLE (two meetings a year), members expressed their expectations, work was organised in four groups (river, canal, marsh and water table), and the Permanent Commission (bureau) met fairly regularly. CLE also formed partnership agreements with the PNR and the Water Agency, which funded 65% of the cost of producing the SAGE document.

But CLE's work has stopped between the end of 1998 and 2000 because the Park representative responsible for leading the CLE resigned her post. With the arrival of the new team, work was reorganised and reactivated. During 2000, two studies were launched, one concerning flood and water level issues and the other water flow in the basin. The purpose of the first was to tackle the problem of run-off and soil erosion, while the second was a response to the need for quantitative management of the water resource, defined following consultation with all the CLE partners. A number of meetings were held with elected representatives to inform them and raise their awareness of the problems of flooding, run-off and erosion. These two studies led to documents outlining the diagnosis and issues at stake. The results of these studies will allow conclusions to be drawn which should lead

[3] Due to the novelty of the SAGE procedure (the Audomarois initiative is one of the first in France) the administrative procedure of boundary-definition and composition of the CLE actually took three years.

[4] The State representative in the *Département*.

to the formulation of strategies to be implemented in the framework of the final SAGE document. This phase had not yet been achieved in 2001.

The SAGE process appears to be a major but time consuming step for promoting the integrated management of the water resource in Audomarois. Four main conflicts mostly dictate water management in the Audomarois and therefore influence the content and the decision process of the SAGE. These conflicts are:

- industry versus flora & fauna (problem of quality) ;
- navigation versus market gardening (problem of quantity) ;
- agriculture and market gardening versus flora & fauna (problem of quality) ;
- drinking and industrial water withdrawals versus flora & fauna (problem of quantity).

The evolution of these conflicts is described below in order to emphasize the reasons why water resource is (partly) managed in the Audomarois.

5.3.2 Management of conflict between industry and flora & fauna

The problem concerns particularly the alteration of water quality due to industrial effluent. Eleven establishments constitute the industrial fabric of the Audomarois. There are six paper-cardboard manufacturers, a vegetable conservation plant (Bonduelle), a crystal glasswork manufacturer, two other agro-food companies, and a cement factory. All these industrial establishments come under the regime of establishments listed for the protection of the environment (ICPE = *Installations classées pour la protection de l'environnement*), which are subject to authorisation. Only the State has powers concerning the legislation on listed establishments. Industrial waste therefore comes under ICPE regulations defined through prefectoral orders, which contain specific measures applicable to each company. They are subject to norms that each company must respect, and which are monitored by inspectors. The right of these companies to pollute water is therefore determined and restricted by public authorities.

Up to the sixties, industrial development in the Audomarois was limited. However, effluent from such plants as these were, was poured straight into the environment (the Aa), as there was no provision for waste-water treatment. During the second half of the sixties, industrial activity in the Audomarois increased, with increasing pressure on water resources. This pattern was matched by an increase in effluents. The growth of industry led to a deterioration of the environment: the environmental quality of the marsh was threatened by polluted water and silting-up of the canals, leading to a decrease (or even disappearance) of the most sensitive aquatic flora. At the same time, eutrophication radically altered aquatic life. These impacts also brought about a reduction in fish life. This situation was the source of a conflict between industrial activities and fishing. At the same time, the 'nature protection' use, the existence of which had not generally been taken into account up till then, began to develop. Regulations, together with public opinion (becoming increasingly sensitive to the environmental issue), led to growing awareness of industrial pollution. In the seventies, paper mills located in the lower Aa valley installed water treatment plants but the improvements achieved were temporary.

This latent conflict finally exploded with a pollution incident at the Bonduelle vegetable-processing plant in 1987. All the actors and water users (fishermen, riverside residents, nature protection organisations, the Park, etc.) reacted in concert. The Park took advantage of the scale of the disaster to raise the awareness of water users of the need to manage the water resource. At the suggestion of the Water Agency and the Park team, the Préfet of the region decided to set up a Consultative Committee, bringing together industrialists, nature protection associations, State departments, local authorities, etc. For the next four years, this committee met regularly twice a year. After initial reticence to invest in anti-pollution measures, Bonduelle made some improvements to its discharge system. However, most of the committee members considered these to be inadequate. Another incident in November 1990 forced the company to undertake the construction of a new waste water treatment plant, with financial assistance from the Water Agency. In 1990, it was decided to include paper mills in the committee meetings, the water treatment measures they had implemented in the seventies having become inoperative. The issue, which had initially been limited to the marsh and to the Bonduelle firm, was thus extended to cover the whole drainage basin. Moreover, the local authorities were not left out, and a sewage treatment plant was put into service in October 1990 to treat the sewage of the communes of the Saint-Omer district, with financial assistance from the Water Agency. Altogether, five water treatment plants (for *Bonduelle*, certain paper mills and the city of Saint-Omer) were created following these meetings, and these have largely reduced water pollution in the area. The urgency of the situation also led to setting up a system for measuring water pollution, mostly funded by the Water Agency, which further influenced the behaviour of industrial firms. Since then, the volume of discharge authorised in the Audomarois is generally respected and industrial pollution no longer appears to be a major problem.

The new water law (January 1992) provided the Park with a regulatory national framework for local actions to preserve water quality (thanks to the SAGE procedure). It opened up the possibility for global management of the water resource throughout the drainage basin, continuing the consultation process between different water users, which had been started in 1988. But the procedure accompanying the drawing up of the SAGE, however, appears only weakly relevant here because corporate management of DRIRE-listed factories resumes its normal course once the anti-pollution plants have been built.

So, as the sources of industrial pollution are fairly easy to identify, action has been put into effect rapidly. This action has been based on financial incentives and compliance with discharge norms within the framework of a collective based procedure. The emergence of a consultative procedure involving a large proportion of water users in the Audomarois following the incident at the Bonduelle plant set the framework for the implementation of these measures. This consultation procedure also trained local actors to act together and facilitated the SAGE process. Management of this conflict was made possible essentially by two key actors who have intervened practically (the Park, with coordination resource, and the Water Agency with financial resource) and on a combination of factors (urgency following a series of incidents, existence of closely-knit voluntary organisations who demonstrated their frustration to elected representatives, the media and the préfet). The novelty of this movement initiated in the second half of the eighties lies in the participation by many users and administrative bodies in defining industrial anti-pollution policy in the Audomarois, a policy which had previously been limited to industrialists and the DRIRE. As a matter of fact, this movement reflects more 'internal coherence' of the water public governance system, if we refer to the categories used within the Euwareness project.

5.3.3 Management of conflict between navigation and market gardening

Navigation on the Neufossé canal is a major use (for leisure and for industrial transport). Industrial transport requires a high water level in the canal. Navigation being a major economic activity for the canal manager (Voies Navigables de France -VNF- which is paid in part by the dues paid by the boats using it), it is this that primarily determines the water level in the canal. Up to the construction of the Neufossé canal between 1958 and 1965, there was no major cohabitation difficulty between market gardening and navigation uses. Variations in water level were not a particular problem for market gardening, as on the one hand in winter when levels varied considerably there was virtually no market gardening activity, and on the other, in summer there was little variation because of the limited needs of navigation and it was not affected.

From the 1960s, the market gardeners specialised and produced two crops instead of one per year. Therefore, flooding of the marsh-land in winter was a problem, as winter crops require a low, constant water level. The demands of intensive agriculture also involve abandoning traditional methods of water transport in favour of motorised tractors, which need tracks to be created for access to the plots. Meanwhile trends in navigation demanded a high water level in the Neufossé canal (and consequently in the marsh) for bigger boats to be able to use the enlarged canal. From the end of the sixties then, market gardening and navigation have been conflicting uses in terms of water-level.

In the absence of rights and with their new demands/constraints regarding water level, however, market gardeners' use of water was protected in two ways. On the one hand, they set up a system of closed fields (by dikes) in the marsh inside which the water level is kept constant by pumping. This system allowed market gardening to expand, protected from the risks of the natural variation of water level in the marsh. On the other hand, market gardening has benefited from measures taken to evacuate water, particularly flood water. Indeed, the catastrophic floods of 1974 and 1976 led to the creation in 1977 of an Interdépartementale institution to set up flood-water drainage systems for the *wateringues* area (*Institution Interdépartementale des Wateringues* -IIW) on the initiative of the General Councils of the *Nord* and *Pas-de-Calais departments*. IIW has been responsible for installing, managing and maintaining eleven flood-water evacuation stations, diverting water to the sea via the Neufossé canal. Market gardeners have regularly lobbied their elected representatives to intervene on their behalf with the Préfet or directly with the managers of the canal structures that regulate water level (IIW and VNF). But VNF derives its income almost exclusively from taxes paid by boats using the public waterway system. It is easy to understand, then, that commercial navigation is an economic priority for VNF in the Neufossé canal while also taking market gardening uses into account.

Faced with the reality of this situation, management in the eighties and nineties appeared to be a matter of compromise. In 1980, by mutual agreement (a verbal agreement only, with no statutory power) with the market garden profession, VNF guaranteed a maximum level of water in winter. In 1995, a 'Protocol of management of the water of the River Aa and the large canal' was drawn up and signed by the different actors concerned by the quantitative management of the water resource, stipulating a new maximum water level. Thus, in order to respect this agreement, pumping and sluice-gate operations by IIW are required even during periods of no flooding, and these are expensive. Currently, within the framework of the SAGE, a specific study is being carried out into 'management of flooding of the Aa and water levels in the marsh'. Working group meetings have allowed communication between the two parties to improve. In September 2000 they agreed to set up a system of three measurement points in the marsh giving the inhabitants (and thus the market gardeners) points of reference.

So the creation of the IIW reflects the need for an organisation which is at the same time the beneficiary of funding from a number of area authorities (the Region and its two *Départements*) and the contracting authority for the operations to be carried out, and also has a special relationship with both public authorities and users (market gardeners and the State navigation service of the *Département*). It provides the technical means through which these users can reach common agreement and mutual acceptance. Successive agreements provide evidence of a *real involvement of dominant users and the governing bodies concerned,* with the absence of other users or their defenders (e.g. the Park for nature conservation). We can see here an attempt at a more coherent water use management i.e. a *better coherence of property and use rights system* (to refer to the categories of the research project). However, the SAGE procedure begs the question about the openness of this community to other uses, which might lead to a better external coherence i.e. a better integration between the public governance system and users and the property rights system. Generally, though, it is the local dialogue that has led to management, albeit limited, of this user conflict.

5.3.4 Management of conflict between agriculture & market gardening and flora & fauna

In the lower Aa valley, agricultural activity is characterised by large, mainly cereal farms (with a total surface area of about 14,000 ha), while in the upper Aa basin it is mainly cattle farming, often conducted by the establishments listed for the protection of the environment (ICPE regime). In the marsh, agriculture is dominated by market gardening, which is the traditional activity in the marsh although today it covers only 15% of its total surface area. For market gardeners, whereas their user rights to water are closely linked to their status as riparian owners,[5] that make them responsible for upkeep of the watercourse banks and bed to facilitate the flow of water. Their rights and obligations are more specifically defined in the rules of the 7th *Wateringues* section, applying to all landowners of the marsh since 1856 (i.e. about 10,000 landowners) that aims at keeping the *watergangs* in good

[5] Article 644 of the Civil Code and a law of 1807.

condition. Up to the sixties, it was essentially the market gardeners who shaped and preserved the marsh through their activities.

The wetlands of the marsh are an ideal habitat for a number of rare and endangered species of bird life and are recognized by various quality labels from the State and the *Département*, which has pursued a policy of purchasing and protecting sensitive natural areas. The quality of surface water in the marsh is therefore an important issue in protecting this environment.

Changes in farming and market gardening in the sixties led to increased pollution (due to discharge), which had a serious impact on the natural environment. Less use of natural fertilisers and their replacement by more profitable chemicals, together with intensification, polluted first the Aa and its tributaries and then the marsh water: hedge removal resulted in leaching of the soil, with large quantities of sediment and chemical fertilisers being carried to the river, increasing turbidity with a total destruction of aquatic plants and sedimentation which clogged the canal and the *watergangs* downstream, but also erosion of banks and increase of suspended load.

More specifically in the marsh, there was an increasing lack of maintenance of the *watergangs* because the number of market gardeners decreased (by 50% in 20 years) and chemical fertilisers replaced mud from the canals. As a result, the canals became clogged, preventing the natural flow of water. This situation was aggravated in the seventies by the construction of roads that were built for the economic development of this area and especially for opening up market gardening activities after a crisis following severe flooding which destroyed a large part of the crops in 1974 and 1976. The negative impacts of this economic development policy promoted by the Saint-Omer District were obstruction the flow of water and decreasing of the market gardeners' incentive to maintain the *watergangs* because it was easier to use these roads rather than the waterways. Thus they suffered both from the consequences of their own actions and also those of agricultural activities in the Aa valley. Very soon, the 7th Section was unable to take the place of the market gardeners to see to the upkeep of canals due to lack of means which led to the deterioration of the marsh, its flora and fauna.

To make up for the disengagement of the market gardeners and the deficiencies of riparian owners, the 7th Section adopted stronger methods of action with help from the Park, which carried out an ethnological study on the market garden population and funded maintenance equipment and technical and financial assistance from Voies Navigables de France (VNF;

National French waterways).[6] The Park also appointed a team to carry out restoration and maintenance operations for the banks and bed of the Aa. Some communes undertook works in the common interest to reduce run-off from agricultural land (hedge-planting, dyke-construction, etc.). For its part, the General Council stopped financing certain operations, which had a negative impact on the environment (drainage works, etc.). At the end of the nineties, a collective Regional Production Contract (CTE = *Contrat Territorial d'Exploitation*)[7] was also initiated by the Park to tackle soil erosion by helping farmers control run-off as far upstream as possible.

Significant improvements have been observed in this conflict, but the situation is still far from providing an example of integrated (coordinated) water management. Individual rights have taken precedence over obligations, which are in fact partly subject to those rights. State action has been almost non-existent. However, the coordination of actions to restore the balance between market gardening and nature is still in its early stages in spite of all the efforts, which have been taken. It is only in the last few years that users have been involved in managing these problems in the framework of the SAGE procedure so that the *extent* of the damage they are inflicting on nature can be taken into account. We can question today whether market gardeners and farmers are really interested in being involved in the policy undertaken by the SAGE in particular. The Park is now acknowledged locally as having the right to act and as being the driving force in tackling this conflict with long-term action.

5.3.5 Management of conflict between drinking & industrial water withdrawals and flora & fauna

Given the poor quality of surface water, drinking water supply comes from bore-holes in the water table. There are four main catchment areas in Audomarois (2 million m^3 water per year for the smallest to 20 million m^3 for the biggest) managed by private or semi-public firms, all on behalf of inter-municipality structures gathering communes mostly outside the Audomarois and for which water is thus exported.

All groundwater abstractions have been subject to the regime of authorisation (more than 80 m^3/hour abstracted) since the 1992 water law. The protection boundary for the catchments area is the subject of a prefectoral order that defines the volume which can be abstracted, and the

[6] VNF; National French waterways is a public organization partly independent of the State that is responsible for management of public watercourses and therefore for the Neufossé canal.

[7] Law on agricultural development of 9 July 1999.

compensation to be paid to affected land owners and users, following a public enquiry. Abstractions are monitored jointly by the Water Agency (for the volume) and the Departemental State administration for health and social affairs (for water quality conformity).

As far as water consumption by industrial firms is concerned, it originated mainly from the water table and four companies that pump surface water (two from the Aa and two from the canal). The regulations that provide the framework for industrial withdrawals come under the auspices of the Regional State administration for research, industry and environment (DRIRE) without real coordination with other water policing authorities.

Up to approximately the end of the sixties, no reference had been made to any possible over-use of groundwater. But the abusive use of this resource has an inevitable impact on animal and plant species in the marsh. It was towards the end of the sixties that the over-use of the Audomarois water table was revealed because of pressure from industrial development in the Dunkerque region (with its corollary of high water demands) and an increase in the population of the Lille region. At a time of rising environmental awareness, a conflict emerged through over-use of groundwater, between on the one hand demand for drinking water in regions outside the Audomarois and industrial water needs, and the protection of the environment on the other. Thus, during the first half of the seventies, one of the objectives in the multi-year plan of the Water Agency was to conserve the groundwater reserve.

The industrial impact on water abstraction has decreased globally by 25% since the 1970s (18.3 million m^3 in 1972 to 14 million m^3 in 1992) and is now stable (about 12 million m^3 in 2000), while at the same time there has been a major transfer of water abstraction from groundwater (a reduction of 45% between 1972 and 1992) to the River Aa and the Neufossé canal. Improvements in recycling and treating water have also given a new direction to water consumption by companies.

However, it is the supply of drinking water that causes the main uncertainty concerning the control of abstractions. After a rapid rise in the seventies, demands for drinking water are generally stable today, although they rose from 21 million m^3 in 1972 to 31 million m^3 in 1992, an increase of 50% in 20 years. This 'stabilization' resulted from a dual action by the public authorities:

- First, by setting up protection perimeters round the catchments areas. The 1964 water law took up the Public Health Code and made provision for the creation of protection perimeters round catchment areas supplying local communities, following a procedure of declaration of public good (DUP = *Déclaration d'Utilité Publique*, a compulsory procedure under the 1992 water law). However, 80% of wells for drinking water in the Audomarois have not yet implemented this procedure, revealing an unwillingness on the part of the local authorities because these perimeter zones result in a restriction of land use, particularly for housing developments or economic activities. This shows the extent to which local authorities control the implementation of public policy through their powers in allocating land use and thus indirectly user rights (through drawing up town-planning documents). The corollary to these rights is a fragmentation of abstraction uses, which makes any desire to control the issue a delicate matter.

– Second by limiting the quantities abstracted: authorisation for new withdrawals is dealt with by a procedure involving a State department and the water company.

Furthermore, within the framework of drawing up the SAGE in the past years, groups of experts have met (in preparation for a 'Water-table management' working group) to discuss the quantitative management of the resource. This conflict of uses seems actually to have been taken into account and integrated into future regulations in the SAGE document.

However, the development of this conflict is not very clear. Still today, in spite of the efforts which have been made to limit withdrawals of the water resource by local authorities and industrial plants, the situation does not seem to have stabilized. Public policy based both on regulation and financial incentives has clearly been the main factor in limiting the withdrawal of the water resource. However, user rights for abstracting water (granted by prefectoral orders) and land use rights appear to be *the main obstacle to implementing a coordinated and coherent policy* for managing the groundwater resource in the Audomarois. The use and protection of this resource seem to be so important that they have remained the preserve of protagonists who pre-date the creation of the Park (DRIRE, Water Agency, business managers and authorities mostly located outside Audomarois). The Park is not so far involved in decision-making on matters concerning water abstraction, the management of which remains relatively confidential and involves no form of coordination between managers and between them and other users. Management of this user conflict remains complex in so far as water withdrawals continue to depend almost entirely on sectoral activity and on stakes that go beyond the regional boundaries of the Audomarois. It seems, anyway, that there has been little progress in discussions on the issue within the SAGE framework.

5.3.6 On the way to an integrated regime in Audomarois Basin?

A movement to coordinate conflicting water uses in Audomarois has led to what can be called an *ongoing integrated regime*. This was initiated by the creation of the Park at the end of the seventies (which reflected a regional will to implement an integrated development policy in this region) and the creation of the consultation committee in the eighties and then by the process of drawing up the SAGE in the nineties, a document which should eventually lead to coordination of the various uses. This is a gradual but slow process.

These three initiatives show that there was a local will, fairly innovative for the time, to implement global management for heritage and environmental problems at the level of the drainage basin as a whole. The Water Agency contributed significantly to this trend through its financial support and the Park through the lead it gave to action to be taken with the Water Agency's support. The novelty of this process lies in consideration of the whole Audomarois basin, while previous efforts at coordination were aimed essentially at the marsh. However, while SAGE enables representatives of the various users to meet (in the CLE and different working groups) without any tangible result so far, there is involvement and partial coordination of uses.

These experiences actually showed the difficulty of maintaining administrative coordination over time, considering the SAGE has still not been put into effect after more than seven years of procedures. It also raises the question whether this coordination only concerns the least serious issues. If it is true that the four conflicts studied are implicit in the thrust of the preparatory documents of the SAGE, it is equally true that (i) the question of water level (navigation v. market gardening conflict) continues to be dealt within a limited community of users and administrative bodies, (ii) the question of industrial pollution (pollution v. nature conflict) continues to be the preserve of the DRIRE, (iii) the crucial question of abstracting water for industrial and domestic use (abstraction v. nature conflict) continues to be the preserve of a few local authorities, companies and devolved State departments without any coordination, and (iv) agricultural pollution (agriculture and market gardening v. nature conflict), has only been the subject of concerted action in the last two years.

Thus the development of each of the four conflicts studied has seen the constitution of a restricted community, usually consisting of representatives of the directly affected users. Such parallel movements can be understood as an attempt at better coherence on the one hand at the level of public governance system and on the other hand at the level of the property and users rights system. The external coherence between both systems seems to be on the way.

5.4 The Sèvre Nantaise basin

5.4.1 The Sèvre Nantaise drainage basin and main water uses

The Sèvre Nantaise is a tributary of the Loire, with its confluence at Nantes. The river and its drainage basin are called *Bassin Versant de la Sèvre Nantaise* (BVSN). The BVSN straddles two regions, four *départements* and 115 *communes*, with a surface area of 2,493 km² (See Figure 5.2).

Figure 5.2: *The position of the Sèvre Nantaise basin in France*

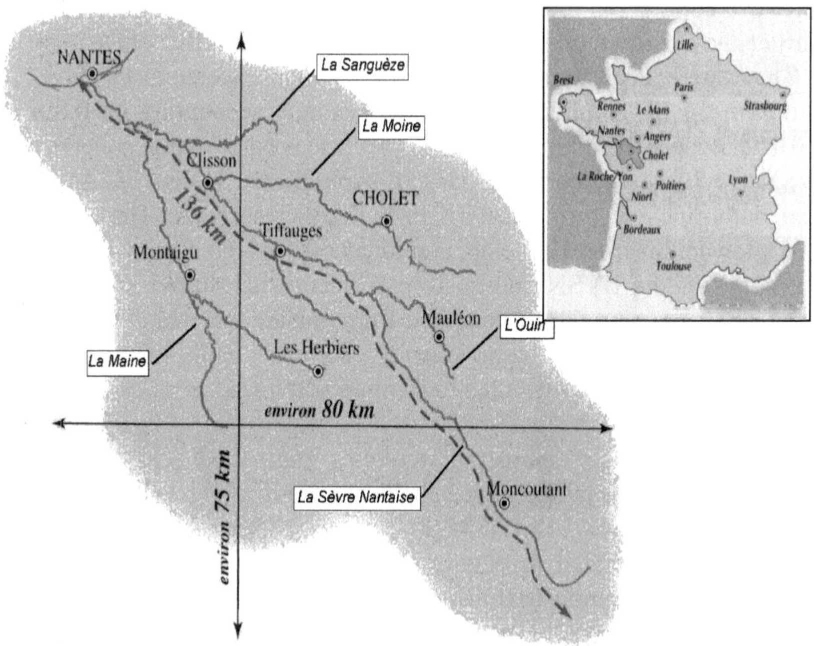

The south (upstream) of the BVSN is characterized by a *bocage* (hedgerow) landscape and mainly agricultural activities. To the north (downstream), farmland is gradually replaced by the vineyards of the Nantes area, and by industrial and urban districts.[8] The BVSN is a predominantly rural area with Cholet as the largest town (69,024 inhabitants in 1999). The Sèvre Nantaise

8 With a high density of population for the Nantes agglomeration which borders the BVSN (566,750 inhabitants in 1982, 674,115 in 1999; population not included in the BVSN).

is 136 km long, and before reaching the Loire it is fed by a network of tributaries, the main ones being (from upstream down) the Ouin (33 km long, high water discharge 3 m^3/s), the Moine (62 km long, high water discharge 14.3 m^3/s), the Maine (95 km long, high water discharge 14.1 m^3/s) and the Sanguèze (40 km long, high water discharge 5.31 m^3/s).

The flow pattern of these watercourses is very irregular and occasionally torrential. Winter floods can have damaging consequences on economic activities and residential areas. In summer, discharge can be very low due to the poor specific retention, which leads to minimal flow (or even drying up). To satisfy the needs of human activities, there is a long history of construction work along these watercourses, 235 such structures being identified in 2000. For the most part these are mill dykes (small dams) built to supply water for the flour mills, electricity production, tanneries, abattoirs, etc., or sluice dams to regulate the flow of water. In order to understand water use conflicts in the BVSN and which management processes of these conflicts have been set up, water uses are presented briefly below.

The watercourses supply most of the available water in the BVSN (21 million m^3 in 1999), underground resources only accounting for 0.8 million m^3! The issue of drinking water supply has been a major concern in the BVSN for 30 years due to the lack of groundwater resources. As a result, local authorities invested heavily in building reservoirs. These structures have allowed this use to develop, and today 50% of the drinking water of the communes of the BVSN comes from surface water stored in the drainage basin. The remaining 50% is supplied from outside the area. Today, some of these reservoirs are experiencing serious pollution, particularly eutrophication, mainly attributed to farming activities.

Water abstraction for industrial activities (agri-food; clothing: tanning, shoes, textiles; metallurgy, mechanical engineering, tyre and paper industries) is relatively low (895,000 m^3, i.e. 7.5% of total abstraction in 1997) and concentrated essentially in the Cholet area in the east of the BVSN. In 1997, 96 industrial plants in the BVSN were liable to taxation for water abstraction (mainly from stream-fed reservoirs, direct withdrawals from natural watercourses, from deep wells and from the alluvial sheet).

Abstraction for irrigation purposes represents a significant proportion of water needs (12 million m^3 abstracted in 1998, i.e. approx. 57% of total abstraction in the BVSN) whereas only 5% of the agricultural land of the BVSN is irrigated. The BVSN as a whole consists largely of farmland, the main activities being extensive cattle farming and intensive (soil-less) pig, chicken and rabbit farming. There is also increasing cereal cultivation, particularly in extreme upstream areas. Wine-growing (Muscadet) is a major economic activity downstream, and there is some arboriculture upstream.

Farming activities as a whole are also the source of major pollution from livestock waste and increasing use of fertilizers and pesticides. But there is also household sewage disposal that can cause pollution due to malfunctioning or inadequate sewage treatment systems for private households (diffuse pollution) and local communities (inadequacy of some sewage treatment plants for current needs, or one-off polluting incidents which interfere with their normal functioning). However, on the whole, household sewage treatment has improved significantly since the eighties. Industrial waste creates a problem of overloading the mains sewerage system, causing malfunctioning of sewage treatment plants which in turn leads to watercourse pollution. This situation has particularly affected the Moine sub-basin due to the industrial activities of the Cholet area. However, discharge norms imposed on industrial plants have improved the situation, encouraging some plants to install pre-treatment facilities for their waste water.

The physical characteristics (low flow in summer, soil impermeability), together with the socio-economic context of the BVSN, result in poor water quality. The most significant features are organic matter, phosphates, nitrates and pesticides. Thus, the quality of the Sèvre Nantaise upstream is considered average, but deteriorates significantly below the confluence with the Moine which carries polluting matter from the industrial plants and local communities of the Cholet area. A strong indicator of the severity of the problem is a ban on swimming in rivers and reservoirs. *Fishing (numerous* associations with 22,000 members in 1998) is a major use on all the drainage basin watercourses that is also affected by this poor water quality and severe low-water periods.

Apart from the very downstream area, the river and its tributaries come under the private domain. Access to the banks is subject to authorisation by their owners, and the watercourses are policed by the Agriculture and Forestry Office (DDAF= *Directions Départementales de l'Agriculture et de la Forêt*) of the four relevant Départements (i.e. four offices). All these different uses are actually regulated by various water policing bodies in the four *Départements*. Consequently, the police departments that intervene may vary from one bank to the other, and there are in fact (too) many different administrative bodies involved in the whole of the BVSN.

5.4.2 The emergence of user conflicts after World War II

From the end of the 19th century to the end of the 2nd World War, the BVSN area remained rural and agricultural. Life was organised round the river, its value being recognised for its usefulness and importance as a means of daily subsistence. Works of common interest were therefore carried out (constructing weir sills, dams and bridges, and cleaning watercourses). In the post-war period, and following the rural exodus, there was a move away from mixed subsistence farming towards increasingly market-orientated practices with new cultivation methods, intensive cattle farming and soil-less farming. While the river had previously been seen as a resource, which needed to be sustained, it rapidly became a tool for intensive production to be exploited as a way of meeting increasing agricultural demands (the law of 1898 allowing riparian owners unlimited abstraction). Some actions had severe consequences on rivers such as increasing run-off and pollution flowing into the river due to hedges pulled out on a large scale because they hindered the movement of agricultural vehicles, or major drainage work (digging ditches and putting in land drains), carried out to make land cultivatable which until then was only fit for grazing. Meanwhile, farmers and people who replaced them (and had no interest in river use) gradually abandoned its upkeep and banks became overgrown and collapsed. Moreover, fuel-oil replaced water power, so activities using this new form of energy no longer needed the structures for flow control and water level maintenance. And so on. All these examples provide evidence of how the BVSN water management regime was becoming more complex.

5.4.3 The foundations for a water management policy in the 1970s

A major issue came to the attention of the national and local authorities at the mid-1970s: the quantitative control of water, particularly, that used for agricultural purposes. With a first study conducted in 1974 by the regional office of the Ministry for Agriculture - the Regional Water Development Department (SRAE - *Service Régional d'Aménagement des Eaux*) that concluded thirteen dammed reservoirs should be built in order to fulfil water needs for agriculture activities, an influential local elected representative (mayor of a *commune* in the Vendée) requested the Préfet of the Pays de la Loire Region in 1976 to designate a coordinator for all studies and actions to be carried out for pollution control and landscape conservation in the BVSN. In 1977, this influential politician became Minister for the Environment. At his request, the SRAE drew up some documents: a global development study of the BVSN based on a set of surveys[9] that led to a framework for a water management policy in the Sèvre Nantaise area, a detailing status of projects for installing domestic and industrial water treatment plants, and two other

9 Analysis of socio-economic development and spatial distribution of activities; under the headings of water works (pollution control, agricultural water management) and tourism; comprehensive development area map.

studies on the potential for developing irrigation in the basin and on the issue of land drainage.

Due to the disparate nature of the administration of the BVSN, the need for a legal body to coordinate actions throughout the BVSN then became clear. Thus in 1978 the Association of the Sèvre Nantaise and its Tributaries (ASNA) was created. The purpose of this inter-*département* organisation was to manage water in the four *Départements* of the BVSN. It brought together 95 communes, four Départements, two regions, chambers of commerce and user groups. The SRAE was responsible for the technical coordination of studies and the water development works initiated by ASNA.

Three new studies were thus carried out at the request of ASNA and provided the foundation for actions undertaken up to now. The first was carried out in 1979, with funding from the Pays de la Loire Region (PDL Region), to look at the problem of low-water periods and to suggest ways that farmers could use abstracted water for irrigation in summer. The study concluded that a dam was needed above the confluence of the Loire (*Pont Rousseau* dam). The second study, carried out in 1981-82 with funding from the Ministry of Agriculture, the PDL Region and the Loire-Bretagne Water Agency (AELB), was a Water Development Plan ('*Schéma d'Aménagement des Eaux*'). This study highlighted the problems concerning water as a whole and enabled essential work to be prioritised: maintaining water levels in summer, and improving domestic and industrial water treatment. A third study -- the first in France -- deriving from the field of research and entirely funded by the Ministry of Agriculture at the request of the SRAE, brought in 'water heritage management'. This study identified the actors and their connection with water and defined the appropriate means to ensure water heritage management in the BVSN: the creation of a decision-making body for water-related issues; the appointment of people responsible for water in each *commune*,coordinated by a technician (river warden) in each sub-basin; and the creation of a Water Centre where user conflicts could be dealt with.

5.4.4 Studies in the 1980s recommend more coordination and water works

The most urgent tasks for the ASNA in the 1980s were to improve water treatment for households and industrial plants not connected to the mains system and to ensure water supply during the summer low water period.
A work programme covering the whole of the BVSN was initiated with the signing by ASNA of a Rivers Contract in 1984 (the sixth in France, after a two-year procedure) depending on a 1981 circular from the Ministry of the Environment. A river committee, co-chaired by the Préfet of the Vendée and the President of ASNA, was set up to ensure the contract was properly

fulfilled up to 1989: installation of sewage treatment plants by local authorities and water purification measures by industrial plants for pollution control (63% of the Rivers Contract budget); repairing and building dykes for river and river-bank development (35%); and landscape protection, improvement of beauty spots and development of tourism to increase the profitability of the river (2.5%).

Implementation of the Rivers Contract led ASNA to question how works for the operation as a whole could be supervised. Because ASNA had no formal power to manage the implementation of operations, it decided to set up an inter-*département* structure with legal and administrative powers and financial means, as recommended in the water heritage management study. The final choice was for an inter-*département* institution of the Sèvre Nantaise Basin (IIBSN) that is created in 1986. It became then the tool of the ASNA for supervising implementation of the Rivers Contract and for dealing with quantitative issues.

In order to meet agricultural demands by which farmers wanted actually to reduce the period during which riverside land was under water and to satisfy their needs for water in summer by irrigation, two kinds of works were undertaken. On the one hand, the aim of some work was to evacuate drainage water by digging ditches and removing river obstructions (dead trees and earth), but also through major watercourse cleaning and riverbed realignment operations. All the more so as considerable grants from the State encouraged this cleaning and drainage work (financing up to 60% of the study phase and 50% of work carried out collectively). Some inter-*communes* structures were created especially to perform these works with help of the State administration for agriculture (DDAF). Realignment works were finally carried out along 38% of the length of the Sèvre Nantaise and its tributaries. On the other hand, action was also required at mills to maintain water levels during low water periods, by repairing dykes and/or building numerous dams with mechanically-operated sluice gates and manually-operated sluices. Actually, the situation on the river during low levels in summer (when it dried up completely in certain places) was such that these works were necessary for all users (including fishermen). But at the end of the 1980s, many operations such as drainage works, together with river cleaning and realignment work had already had serious consequences for flora and fauna, which were destroyed in many places. Moreover, the faster current (due to straightening of the watercourse) also led to a destabilisation of the banks, increasing erosion and deposits of sand. The increased rate of flow, combined with irrigation, also resulted in a significant reduction in the level of the river during the summer, which also had a negative impact on the biological function and did not please the fishermen.

In 1986, as operations planned in the Rivers Contract were gradually being implemented, it was also decided to go ahead with construction of the Pont Rousseau dam at the mouth of the Loire, as proposed in 1979. However, due to the potential repercussions of this construction work on the flow pattern of the river, the Ministry of the Environment requested ASNA to revise the Water Development Plan to include its impact. Other local authorities also asked for dam projects (for water supply or irrigation) to be taken into account. However, in the revised Water Development Plan, the SRAE expressed reservations about the advisability of these projects due to the large amounts of phosphates produced by farming activities in the water, creating the risk of making it unusable for drinking water or of causing eutrophication. Nevertheless, these construction works were passed by the IIBSN and included in the revised Water Development Plan in 1988. Construction work on the Pont Rousseau dam was started in 1990. This was a major project for ASNA and was completed in 1995. Two other dams were built in the 1990s.

But a big stake for IIBSN was also to create or reinforce local actors for implementation of its policy on the ground, which is why it has developed a global river management policy based on creation or reinforcement of inter-*commune* water associations and on the appointment of river wardens. Inter-*commune* water associations are river authorities with a board of elected representatives from concerned communes that decide actions to be implemented for river management. The river warden's job is to carry out specific river management and maintenance operations and to liaise with the river authorities and IIBSN. This enables action throughout the BVSN to be coordinated. The IIBSN covered 75% of the administrative cost of these river wardens, and put them at the service of the inter-*commune* water development associations. In 1988, the IIBSN recruited the first two river wardens who would be involved with the two existing river authorities. Their first tasks were to carry out a survey of the rivers and prepare a multi-year programme of work. However, not all the BVSN was covered by the river authorities. The IIBSN therefore started discussions with the local authorities of these sectors in order to encourage the creation of inter-*commune* associations, which has led to the creation of such organizations in 1991, 1993, 1995 and 1999. A river warden was appointed for every new inter-*commune* association. Moreover, another IIBSN policy tool was an 'intercommunity charter for watercourse maintenance', in effect since 1991. This was a contract between IIBSN and the new sub-basin river authority through which the latter would respect the following principles: *"to conserve watercourse flow, preserve its quality by encouraging self-purification phenomena, reduce and spread over time the cost (rather than wait for expensive restoration work every 15-20 years), maintain the biological capacity of the living environment, manage the landscape to encourage leisure activities"*. This provided an innovative system for restoring and preserving the 'biological function' use. Acceptance of this charter would facilitate the creation of posts for river wardens (75 % of total cost paid by IIBSN), whose first work was to carry out an appraisal and then draw up a multi-year plan of restoration and maintenance work of riverbanks and dykes.

5.4.5 Towards greater coordination in water management policy in the 1990s

Meanwhile, in order to consolidate the first actions conducted by river wardens and the SAGE procedure (see below), ASNA engaged first in a landscape protection policy (1993), drawing up a 'Landscape Plan of the Sèvre Nantaise' and a Regional Water-Related Landscape Development Agreement (CRAPE – '*Convention Régionale d'Amélioration des Paysages lies à l'Eau'*), initiated by both the two Regions. Strengthened by specific contracts in the *départements,* CRAPE enabled the IIBSN to coordinate about 50 restoration projects by the rivers authorities, including river bed and bank maintenance, restoration or creation of footbridges, footpaths and landscape improvement work. CRAPE contributed 20% of the cost of river upkeep for three years. Thereafter, in 1997, a 'Restoration Upkeep Contract' (CRE –'Contrat Restauration Entretien') proposed by the AELB enabled restoration and management operations to be continued by each river authority. The CRE enabled action and work by all the rivers authorities to be coordinated for a period of five years. An appraisal was carried out between 1998 and the end of 1999. In February 2000 a work programme was proposed at a total cost of 56 million francs. This contract was signed in 2001 by the seven river authorities (through the IIBSN) for five years, and covered the length of the main watercourses.

Thanks to river warden's actions and to solve problems stemming from works conducted in the 1970s and 1980s, some negotiated solutions were also introduced to minimise the conflicts related to low water levels in summer. One of them is a system for controlling water levels in low water periods that have been developed for use by irrigators in the upstream area. Instructional measuring scales were installed at the twelve small dams and dykes along the watercourses. Three different coloured levels (green, orange and red), representing the levels to be respected in each reservoir, were negotiated at the beginning of the nineties at meetings organised at each dam by the river technician with the irrigating farmers, fishermen and fishing wardens, and the DDAF. This was the first time that collective water level management had been negotiated by consensus in this area. Every week in summer a prefectoral order authorises or restricts withdrawals according to the existing water level for each dam.

But the main recent issue for improvement of coordination in the whole BVSN is achieving a so-called water management and development plan (SAGE). The 1992 water law, which included many of the operations already implemented in the BVSN by the IIBSN, indeed enabled the previous policy to continue by setting up a planning tool for water uses (SAGE) and a local water commission (CLE - *Commission Locale de l'Eau*) that is responsible especially for management of the SAGE procedure. This extended the action of the River Committee that was set up in 1984 with the Rivers Contract.

Even though ASNA made a request for the implementation of a SAGE procedure in the BVSN in August 1992, the State agreement was very slow and therefore the BVSN SAGE project was accepted in June 1994 and the 92 members of the local water commission (CLE) who are representatives of administrative bodies, local authorities and users, together had their first meeting not before September 1997. A SAGE facilitator was appointed by the IIBSN. The first stage consisted of making an accurate survey, carried out by a planning office with help of a SAGE technical group and sub-basin working groups (allowing involvement of local actors) and ratified by the CLE in 2000. After that, different scenarios for how the water resource could be developed were put forward to the members of the CLE, who since the end of 2000 have met in special-interest groups (agriculture, water works, local authorities, industry, natural environment) to give their opinion on the proposals. The SAGE document has not yet been completed. The survey and global appraisal have been completed and the study is scheduled to finish at the beginning of 2002. The BVSN will then have a coherent water management tool that should allow it to harmonize action throughout the area. The SAGE document should then be ratified by different authorities in order to bring into effect the measures, which it advocates for the BVSN as a whole. The SAGE will put into effect the final recommendation proposed by the water heritage management study of 1983: the creation of a single regulation for water throughout the BVSN. It took twenty years to achieve all that.

5.4.6 Factors framing the implementation of an integrated regime in Sèvre Nantaise

One cannot sufficiently stress the major role played by the administrative coordination of the different geographical scales of intervention. This coordination contributed to the development of global river management, and could be seen in particular: (1) in studies that were carried out; in (2) the creation of coordinating structures (rivers authorities, ASNA and IIBSN); and in (3) the appointment of river wardens. The SAGE constituted an important step in consolidating this administrative coordination (4). However, the policy that has been conducted in the BVSN in the past thirty years is far from being fully completed yet (5) especially because the main issues (i.e. the case of the pollution due to agricultural activities) are not fully addressed (6).

(1) This policy was initially based on the results of three studies carried out at the BVSN level. On the one hand, these results contributed to the emergence of a global view of the problems and issues surrounding the question of managing watercourses throughout the drainage basin, and, on the other hand, they led to the launch of a strategy for action at the basin level.

(2) The emergence of a global view and coherent action at BVSN level was in fact closely linked to a dual level of action: that of the river authority in each sub-basin, and ASNA and the IIBSN at the BVSN level. The advantage of this configuration lay on the one hand in the implementation of an overall management of the resource at each of these levels, and on the other, in the complementary character of these two levels of intervention, which assisted each other. As such, for each sub-basin, IIBSN lay down the principle of creating a river authority as a precondition for its involvement, this authority being both the intermediary for the IIBSN (notably through the river technician, whom it largely paid) and its action tool.

(3) However, this intermediary function of the river authorities depended above all on the intervention of the river technicians. The river technician has worked to raise the awareness of elected representatives, users and riparian owners, particularly those with water-control structures on their land. In this way they are made responsible and consequently involved in developing a balanced management of the river and its uses. In the early nineties, there was thus a gradual trend towards ‘softer’, long-term actions focusing on the regular upkeep of watercourses and riverbanks and restoration of water-control structures. Coordination between users in some parts of the BVSN was made possible thanks to the confidence that users had in the river technician after several years working on the ground and a gradual build up through dialogue. But there is actually a disparity between sub-basins due mainly to the personality of the technician (and particularly his communication skills) and to the nature of the stakes involved in terms of managing water resources.

(4) Finally, drafting the SAGE enabled another step to be taken in coordinating levels of action, while at the same time revealing the gaps that still existed. Coordination was strengthened by the appointment of a facilitator who ensured this coordination at the IIBSN level between the various actors and the regional levels of action. The CLE provided evidence of a real trend towards mobilising and coordinating the various users, instigated and backed by the IIBSN. In this respect, the SAGE formalized the principle of user coordination as a condition for implementing integrated water management. However, as shown by the difficulty the actors involved experienced in drawing up the SAGE and in formalizing the document, just working together is not enough to set new rules for managing water. However, drafting the CRE objectives document, which was presented as a tool of the SAGE, was particularly useful in bringing together the different river authorities. For IIBSN, one of the main advantages of the CRE was not only to obtain substantial funding over several years which would allow it to implement its policy, but above all the opportunity to bring together the technicians and river authorities within a single procedure because, up till then, there were weak cross-connections between rivers authorities. Through the CRE, the IIBSN therefore took a further step in the coordination of action.

(5) There is therefore a real trend towards integration through a policy of contracts (Rivers contract, charters, and the forthcoming SAGE document) which also has been made possible through coordination of the responsibilities and resources of the various protagonists (State department, AELB, ASNA). But the user coordination procedure has still not been completed. Actually, in spite of the generally positive view of the above points, the coherence of the policy in terms of the relevance of bringing public policy and user rights into line remains doubtful. Recognition of ownership rights, particularly those of owners of water-control structures, was only raised at planning meetings for the SAGE in 2000-01. In short, discussion and negotiations on this point have only been started.

Moreover, the administrative division of the BVSN (four Départements and two regions) makes any policy of water management at the BVSN level intrinsically difficult, since each Département draws up its own policy and has its own ways of operating and its own regulations. The size of the BVSN also complicates intervention in the field of water management because of the difficulty of reconciling the expectations of the numerous actors and of meeting the very different needs of users, due in particular to the wide regional and cultural variations in such an area. In the current context of a lack of coherence in user and policing regulations especially between the four *départements* (*internal coherence*), integration of the water management regime by coordinating the application of measures seems to be attempting the impossible. The mere participation of users (*extent*), as promoted by the SAGE procedure, is not enough for coordination of the water management regime in the BVSN.

This administrative coordination in the BVSN was in fact ensured by ASNA and then IIBSN, which has become recognized as the key water manager in the BVSN. The IIBSN now reviews and approves all river authority projects, and has become the key partner for water management actors (the river authorities) and for government authorities and the AELB. However, its powers are essentially linked to those of the river authorities, i.e. to development of the water system through restoration and maintenance work on watercourses. This issue, which was the least complex and least economically and politically sensitive, received particular attention. The same cannot be said for management of the problem of river pollution.

(6) At the BVSN level, apart from a few studies carried out about 20 years ago and coordination of the river contract by ASNA during the eighties, it would be inappropriate to say there is a policy or incentive for ASNA or the IIBSN to carry out a clean-up programme for water in the BVSN. The technicians for the different river authorities under the auspices of the IIBSN deal very little with the problem of pollution and its reduction at source through actions aimed at local authorities, industrial firms and farmers. Up to now the IIBSN has thus kept a low profile on the issue.

The IIBSN actually has no power to intervene in problems of pollution, other than that to raise awareness. Moreover, certain decisions (concerning sewage treatment plants and drinking water) remain in the hands of local authorities, which continue to act independently of any global view or coordinated management. It seems, therefore, that ASNA and IIBSN are 'allowed' to intervene either on the least important issues or where there is space for action (because of the lack of importance of the issues). These difficulties are linked not only to the diffuse nature of the pollution and the complex ways that it is diffused, but also to farmers' power of opposition, derived from their control of land tenure and their corporatism. In other words, the 'governance system' (public policies) is held back by the 'regulative system' (use and property rights).

5.5 Learning from the French cases

The ongoing process of water regime integration in the two cases presents some similarities.

First of all, both cases demonstrate an attempt at *expanding the extent of water uses* taking into account within the water resource regime: environmental issues have been taken into consideration in each river basin since the 80s. The legal framework (water act of 1964, 1984 and 1992) as well as public awareness (linked with problem pressure) supported such an opening. In both cases the recovery of the quality of the watercourse has been linked with nature preservation and 'ecological uses' of the water. When this expansion of uses is taken into consideration, it leads to a more complex regime in both cases.

Secondly, the two cases show an important change toward *a better 'internal coherence' of water public governance*. Regarding this movement, the legal framework is obviously an important explanatory factor. However, such a movement is more obvious for the Sèvre Nantaise case than for Audomarois one. In the Sèvre Nantaise case, the water public governance coherence is forced by the action of the IIBSN. This institution plays an important role in implementing and enforcing this coherence at the level of the whole river basin as well as the level of sub-river basins (with the creation of inter-communal water bodies), by means of involvement of most of the actors concerned within an ad-hoc institutional arrangement. In the Audomarois, such coherence is partially produced by the Park, the boundaries of which do not coincide with the Audomarois basin. In this last case, there are too many actors playing at the same time at different levels, and the Park cannot impose itself as the 'coordination actor' for water policy.

Thirdly, in each case a serious attempt at *a more internal coherence of property and users rights regime* can be noticed. Several agreements have been achieved between different types of water users in order to solve specific conflicts: for instance between market gardeners and VNF in the Audomarois case, and between farmers and fishermen in Sèvre Nantaise one. These agreements are not really set up in the frame of public water policy. They stem from conflict resolution processes, mutual understanding between users involved in the conflict, and they are sometimes facilitated by the implementation of water policy as such.

Lastly, in both cases change towards an *external coherence (between property rights regime and public governance) proved to be very difficult*. The SAGE framework provided some opportunities to enhance such external coherence but not sufficiently: the question of user rights is not really dealt with in the SAGE plans either at the level of the law or at the level of implementation. In other words, in both cases, discussions within the SAGE process have been carried out without paying particular attention to the way these rights have steered and continue to steer past and present uses, and especially without worrying about how well they match the line of action taken by the SAGE, and consequently without paying attention to the obstacles that could arise in implementing such actions. This is the main reason why we concluded our case study report with an evaluation of a partial integration in both cases under study.

However, as concerns the *change agents* leading to this ongoing process of integration, the two local situations differ from one to the other: as a matter of fact, regarding the initiative taken in each case, the Sèvre Nantaise case presents a bottom-up movement whereas the Audomarois one presents a more top-down one.

In the Sevre Nantaise case, one can point out the *role played by the IIBSN as 'water policy entrepreneur'*. Administrative coordination in the BVSN was in fact ensured by ASNA and then IIBSN. Since their creation, these two organisations aimed at developing both the hydrological and administrative identity of the Sèvre Nantaise Drainage Basin. They were able to take opportunities (river contract, SAGE, CRE, etc.) to implement a coherent policy throughout the BVSN. The IIBSN gradually established itself as the coordinating body for a large number of development, management and upkeep projects for the watercourses, and has become recognized as the key water manager in the BVSN. Since the creation of the first river authorities, and particularly since the appointment of river technicians, the IIBSN has followed a policy of encouraging local authorities to come together to form river authorities, with the aim of covering the whole of the BVSN. This objective has been achieved with the recent appointment of the last river technician. IIBSN's main work therefore relies on these river authorities managing one or more watercourses comprehensively, with financial support from the IIBSN and additional funding from the State, the regions, the Départements and the Water Agency. The IIBSN now reviews and approves all river authority projects, and has become the key partner for water management actors (the river authorities) and for government authorities and the Water Agency.

Coordination of actors has been more difficult to achieve for the Park in the Audomarois case. In the Audomarois, the first signs of administrative coordination appeared in the eighties with the creation of the Park (which reflected a regional will to implement an integrated development policy in this region), and then with the setting up of the consultation committee whit the objective of dealing with the problems of pollution in the marsh and which for the first time enabled industrialists and other water users to come together(The functioning and objectives of the consultation committee could be compared to those of a river committee.) These two initiatives show that there was a local will, fairly innovative for the time and still more less divided, to implement global management for heritage and environmental problems. Since the eighties, nature protection organisations and the Park have seen this global management as the only way to deal coherently with the problems of pollution. The creation of the IIW also indicates the awareness of a problem (floods and their impact on market gardening) at the level of an area as a whole (the *Wateringues*) and by users directly concerned by the issue of the water level. Creating these structures or debates enabled a certain number of measures to be implemented, which contributed to managing a part of the user conflicts (carrying out studies which raised awareness of uses and how they interact, expression and acknowledgement of many interests, financial and technical support for the 7th Section of the Wateringues, etc.). The Water Agency contributed significantly to this trend through its financial support and the lead it gave to action to be taken. But these initiatives remain fragmented. The Park did not fully gain a legitimate position of 'water policy coordinator'. Then the implementation of the legal framework plays an important role for opening this institutional arrangement: the SAGE, as a document that should eventually lead to coordination of the various uses, has extended the number of actor to be involved: the novelty of this process lies in consideration of the Aa sub-basin, while previous efforts at coordination were aimed essentially at the marsh (with the exception of operations to install anti-pollution plants in paper-mills in the Aa valley). It was therefore the first time that *communes* of the sub-basins of the Aa and the marsh sat round a table with State departments and all water users. Such an opportunity derived from the implementation of the 1992 Water act.

Knowledge and dissemination of knowledge constitute another important explanatory factor. Many studies have been carried out in each case, on behalf of 'coordinator actors' (IIBSN and Park) and later through the SAGE process. These studies have played an important role to bring about among actors an awareness of interdependencies between water uses.

5.6 Conclusion

France is well known as being one of the first countries to promote river basin water management, of which the early creation of Water Agencies through 1964 Water act is an indicator. The centralized nature of French state administration mainly explains such an innovative policy design. However, property and user rights have not really been changed by this water policy, either by the laws or at the level of their implementation. So, a full coherence of the water resource regime in France has not really been achieved yet. But can we consider that it is on the way? Analysing the concrete implementation of SAGE in the near future will allow an answer to this question. Today we can assess that SAGE processes will be unsuccessful if the user and property rights regime is not addressed. And the poor results achieved by these processes will be underscored with the implementation of Framework Water directive. Then one can expect a movement towards a better coherence of water regime in the near future.

However, the SAGE is part of a movement towards coordinating uses that was initiated in the early eighties. It is true that this movement is slow, but it shows a genuine trend towards global management of the water resource. However, our analysis shows that this coordination depends on an ongoing process that must adapt to changes in uses. In this respect, the SAGE provides answers and a further step towards global management of the water resource. The Audomarois example tends to suggest, however, that the SAGE is not a miracle tool, because it cannot bring about radical change in practices. If it tried to do so, it would give rise to conflicts that would prohibit the achievement of drawing up a document whose orientation resulted from a relative consensus between users, elected representatives and administrative bodies. For that reason national policy support proved to be a necessary, though not a sufficient condition for change. Local institutional arrangement and rearrangement are then a condition to be added, and this is much more dependant on local political will.

Chapter 6

Redistributing Water Uses and Living with Scarcity

The Matarraña and the Mula in Spain

Meritxell Costejà, Nuria Font and Joan Subirats
Universitat Autònoma de Barcelona (Spain)

6.1 Introduction

Water scarcity is probably one of the most salient problems in Spain. The combination of climate conditions, which are typical of the Mediterranean basin, together with the increase of rival uses of water during the last decades, have put water scarcity high on the political agenda. However, while there is a wide consensus on the need to tackle the water issue, the strategies to deal with it seem quite diverse, broadly divided between those in favour of a supply approach and those in favour of a demand one. While policies at the national level seem to favour the first of the two approaches, consisting of the construction of large hydraulic infrastructures to transfer water between river basins, there are a few experiences in which this traditional approach has been replaced by a new one in which the rationality of water use is a guiding principle. The two cases analysed in this chapter, the Matarraña and the Mula river basins, constitute two of these experiences. Departing from a common problem -- serious scarcity -- and following very different types of water regime changes, the outcome in both cases is an attempt to improve the conditions for sustainability. This chapter aims not only to tell these two stories but also to explore the factors accounting for regime change in both cases. The chapter is divided into four sections. The first one briefly describes the national regime context in which the two cases must be understood. The second and the third sections offer an analysis of the two cases, including both the story lines and the characterisation of regime change in each of them. The fourth section analyses regime changes

Hans Bressers and Stefan Kuks (eds.), Integrated governance and water basin management. Conditions for regime change towards sustainability, 165-187.

at the two cases in comparison with national regime changes. The chapter ends with some concluding remarks.

6.2 National context

Compared to other European national regimes, Spanish water resource availability would not necessarily be considered a problem if the available *per capita* water -- which is estimated around 40,000 hm^3/year -- were considered as an indicator. However, it is the Spanish irregularity in water resources distribution, both in terms of time and territory, that makes the difference. Temporary irregularity is, comparatively, more severe in Spain than in other European countries, as annual rainfall variations affect river bed volumes between seasons, and the year-on-year rainfall irregularity results in great differences in river volumes between rainy and dry years. In addition, rainfalls and the demands on water for irrigation and population supply are not coincident throughout the year. Rainfalls and provision levels take place in autumn, winter and spring, while the greatest demands are concentrated in summer time (Pérez Díaz et al., 1996). From a territorial perspective, uneven regional distribution of water resources causes serious deficit situations in the Southeast of the peninsula, and more particularly in the Segura basin, the head of the Guadiana river and some parts of the Júcar, South and Ebro basins.

In this context of water scarcity, whether punctual or structural, the main water use in Spain is irrigation which consumes about 80% of the total amount of water. For instance, in the Ebro river basin, this rate represents 60%, while in the Segura river basin the rate it rises to 90%. The two cases analysed here, the Matarraña and the Mula, are located in the Ebro and Segura river basins, respectively, and reproduce the trends described above: irrigation is the main use of water and scarcity, due both to climate conditions and the increase of water uses along the last decades, is a source of social controversy. While the traditional approach to the scarcity issue has consisted in the regulation of large infrastructures, the two cases analysed in this chapter involve important water regime changes aimed at rationalising the use of water and improving the conditions of sustainability.

6.3 The Matarraña river basin

The Matarraña river is a tributary of the Ebro river on its right side and is located in the Northeast of Spain, flowing through the Aragón, the Catalan and the Valencian Autonomous Communities. It covers 1,727 km^2 and is 97 km long, flowing through twelve municipalities with a total population of 10,613 inhabitants. The climatic patterns of the Matarraña river basin follow the typical Mediterranean regime, which is characterised by low rainfall periods occurring from June to September -- which may cause the river continuity to break down in summertime -- while there are two rainfall peaks in spring and autumn. In addition to intra-annual rain irregularity, important inter-annual rainfall variations can be observed. As a general trend, there are five to ten year periods of scarce rain and drought, followed by five to ten year periods of abundant rain. As regards water uses, the total net water needed in the Matarraña river basin is calculated as 17.029 hm^3 per year, of which 90.4% goes to irrigation, 9.58% is needed for population supply and 0.01% is consumed by cattle.

Even though the Matarraña river basin is neither too long nor too populated, there are important climatic, geographical, social and economic variations in different parts of the river basin. In general terms, three different areas can be distinguished along the river basin: the higher, the middle and the lower basin. The Matarraña river is characterised by being narrow at its head, resembling a torrent or a stream. As a matter of fact, the average rainfall at the head of the river basin is 600 mm per year, while it is lower than 300 mm at the end. The higher river basin has less need of water than the rest of the basin, both because users located there have traditionally had the right to use relatively more water than the rest of the river basin users, and because the extent of irrigated lands is much lower in the higher than in the middle and lower basins. One of the reasons accounting for this is that in the seventies much of the fruit crop that was introduced in the mid-sixties was reconverted into cattle farms, leading to a decrease in water demand. In addition, during the nineties, rural tourism has become an important source of income in this area, but this does not seem to have put too much pressure on the water issue. Compared to the other areas of the river basin, the higher river basin does not have severe problems of water scarcity. However, it has traditionally been the destination of several regulation works that have been promoted by the Ebro Hydrographical Confederation (EHC) in order to secure water for the middle and lower river basins. These regulation works include the Pena water dam constructed in 1930, the diversion tunnels built in 1978, and a water pumping project installed in 1998. Given the perceived lack of efficiency of these works as well as the cost to the higher river basin in terms of environmental impact, expropriations and economic cost, actors located in the higher basin have increasingly rejected projects promoted by the EHC (Arrojo et al., 1997).

The middle river basin has a greater water demand than the higher one, for two reasons. On the one hand, the two main municipalities in the middle river basin, Maella and Mazaleón, represent about 47% of the irrigated land in the entire river basin. This is an important percentage, especially if we consider that irrigation consumes 90% of the total water needed in the Matarraña river basin. On the other hand, during the seventies and mid-eighties, the peach crop, which needs large quantities of water, was introduced in the middle river basin and the familiar production became industrialised. Nowadays, most peach crops are concentrated in the middle river basin. For instance, the municipality of Maella produces around 5.5 million kg fruit every year, of a total of 12 million for the whole river basin. As the extension of the fruit crop through the middle river basin has increased water demands in this area in recent years, the area has become more vulnerable to scarcity problems, especially in dry years. Apart from that, the concentration of most irrigation lands in the middle river basin has had the effect of making users in the middle river basin dominant in the institutions representing users of the river basin, namely the Central Union of Irrigation Communities of the Matarraña river. As they have a dominant position in the river basin, they have greater influence on the decisions taken by the Central Union to confront scarcity problems. This has provoked conflicts with the higher and the lower river basin. With the former, the middle basin has traditionally been in favour of regulating the river by means of hydraulic works that had to be located at the head of the river. With the latter, the middle basin has been accused by users in the lower basin of prioritising the irrigation of lands located in the middle basin.

Finally, the lower basin of the Matarraña river often suffers from severe scarcity problems. Like users in the higher and middle river basin, some users in the lower basin rooted out the traditional dry crops -- mainly olives and vineyards -- and started planting peaches in small areas of land in the late sixties and the seventies. These plantations were enlarged during the seventies and early eighties, but a severe drought in 1986 left Nonaspe lands in the lower basin with no water for seven months. Since then, problems of water scarcity have been constant. Nowadays, due to water shortage problems, many crops have been abandoned or rooted out. It is estimated that in Nonaspe around 60-70% of the land is now uncultivated. In addition, the ageing of population constitutes an additional limit on continuing to cultivate the land. To sum up, two main factors regulate the use of water in the higher, middle and lower river basin: the severity of water scarcity problems and the level of influence on the river basin decision-making processes. These factors are crucial in order to understand the balances and imbalances between actors and uses of the river basin.

6.3.1 Regulative system and *de facto* use rights

At present, the regulative system for water is provided mainly by the Royal Decree Legislative 1/2001, which is legislation that codifies the different modifications of the 29/1985 Water Act. The latter, which is the act in operation during most part of the process analysed in this chapter, declares that all continental waters are hydraulic public domain. While this situation changes the nature of the ownership, the act has had limited impact on the current uses and practices at the Matarraña level in a number of ways, including concessions, the water register and the functions of the EHC and irrigation communities. Regarding concessions, irrigation communities and municipalities, most of which were given their concessions long before the Water Act was adopted, perceive nowadays that concessions are in perpetuity. Regarding water registers, only 12 concessions, which represent a minimal amount of the total water used in the river basin, are currently registered. Finally, in regard to the functions of the EHC and the Central Union, there are also some conflicts between legal aspects and actual practice. The EHC has been assigned certain functions, including the regulation of the Matarraña river basin, although in practice there is a common perception that the Central Union -- and not the EHC -- is the most influential actor in the regulation of the river basin.

Some uses of water in the Matarraña river basin date from centuries ago and are mostly related to irrigation. Most of the irrigation channels still working today were built by the Arab population who settled down in the Peninsula, and the most antique irrigation communities existing today date back to the early XIXth century. Nowadays water for irrigation in the Matarraña river basin can be taken either from the river or from the Pena dam. In case it is taken directly from the river, it is irrigation communities who decide how to distribute water among users. It must be taken into account, however, that the river normally has low quantities of water flowing and therefore the amount of water going into their irrigation channels is lower than the maximum quantity allowed to be used. In fact, the natural river flow of the river does not guarantee having water in the middle and lower basin throughout the year. As a matter of fact, in drought periods, the river is dry during the spring and summer seasons from Mazaleón or Maella. This makes the Pena dam the main source of water to all users in the river basin, but specially to the middle and lower basin. As water from the Pena dam is also insufficient to cover water demands in all the river basin, the decisions on when and how the water from the dam is released become crucial. The EHC is the institution that gives the order to release water from the dam, both in relation to the quantity of water and to the rhythm of the process. However, this decision is taken in accordance with the Central Union, which is considered to be the most influential actor in the decision of how to administer water from the Pena dam. In this respect, it must be taken into account that the Central Union, as a representative body of the irrigation communities, is mostly dominated by those actors having interests in the middle basin, mainly in the municipality of Maella. This village has the largest areas of lands to be irrigated in the river basin and consequently the actors of the middle river basin are dominant in the Central Union: they hold the majority of seats on the board of directors, the president is a person from Maella, and the Central Union office is located in Maella. In addition, Maella has extended the peach crop during the last two decades, and thus water demands in this area are increasingly high. In this circumstance, irrigation communities in the lower basin often complain that water does not always arrive at the lower irrigation channels, while these users pay the same price as the others.

Water is not only used by those having a concession but also by illegal users, that is, by users having no concession. Nowadays, around 650 taps representing at least 30% of the lands in the Matarraña river basin are said to use water illegally: around 1,000 Ha out of a total of estimated 2,300 - 3,300 Ha of the total extension irrigated in the river basin, most of which is located in the middle basin. This situation seems to result mainly for two types reasons. On the one hand, some small vegetable crops, mainly located in the higher river basin, were progressively reconverted into animal farms -- which have a lower demand for water -- a few decades ago. Nowadays, around 60% of the total income in the higher basin comes from cattle. On the other hand, those irrigation communities having no concession have a *de facto* use right given by the Central Union, by which they can take the remaining waters from the Matarraña river from October to March as long as this does not cause trouble to the irrigation communities that have a concession. In turn, illegal users pay for the use of water to their corresponding irrigation communities as a kind of trade-off for being allowed to irrigate, even though they do so in worse conditions than the users holding a concession. Both the dominance of the middle basin actors in the decisions regarding the distribution of water coming from the Pena dam and the existence of illegal users are a source of continuous conflict among users and municipalities. Conflict is specially intense during spring and summer seasons because of water scarcity, which is the period in which illegal users are not allowed to take water from the river but when they need it in greater quantities. The lower basin irrigation communities often feel harmed by the decisions taken by the Central Union, as they consider that the latter mostly satisfies the interests of the middle basin. In addition, illegal users, even though they pay for the use of water, have fewer use rights than the legal ones, but may benefit from the fact of being located in upper parts of the basin, for instance, taking water at night or using water when the lower basin irrigation communities have preference due to their legal status.

6.3.2 The process of regime change

In the early decades of the 20th century the dominant approach to water management was profoundly influenced by the *Regeneracionism* intellectual movement, which favoured State investment in large hydraulic works in order to promote the creation of large areas of irrigated lands and the production of hydroelectric energy for industrial uses. In the Matarraña river basin this perspective resulted in the construction of the Pena dam in 1931. This dam is located in the municipality of Valderrobres, in the higher basin, it has a regulation capacity of 18.5 hm^3 and floods an area of 129 ha. Despite its capacity, it only regulates 8.5 hm^3 per year due to technical problems (Nadal, 1984). The creation of the Central Union of Irrigation Communities of the Matarraña river basin in 1950 changed the water regulation and management system, as this institution centralises powers and is a central actor in the evolution of the basin water regime. During the sixties and seventies new fruit crops were planted along the river basin and the subsequently increased water demand exceeded the regulation capacity of the Pena dam and the Matarraña river flows. In addition, in 1977 there was a severe drought that lasted until 1987. This combination of circumstances led to an increase in social demand for regulation projects within the basin in order to increase the quantity of water stored in the Pena dam. As a response, in 1978 the EHC built a tunnel diverting water from a point higher up the Matarraña river to the Pena dam.

The construction of the tunnel was followed by a severe period which lasted until 1987, which led to a new wave of social demands for regulation projects. Several dam projects were proposed during this period: El Pontet, Torre del Compte and Molí de les Roques, among others. The wet period that followed, which started with the 1988 floods, relaxed the claims for regulation until the beginning of the nineties. In 1993 a new drought started and controversy reappeared. In this context, in I994 the Central Union studied the possibility of pumping water up from a lower point at the Matarraña river to the Pena dam. Finally, the Central Union rejected the project, considering it too expensive. However, the national government approved a new regulation in response to the drought from 1992 to 1995 which provided the legal framework and financial resources to implement the pumping project.[1] This regulation declared that the pumping project on the Matarraña river was of general interest to the State and was considered an urgent measure to palliate the effects of the drought.

The announcement of the initiation of work on the pumping system in July 1995 provoked intense activity in the high basin. The first actors to react against the project were the neighbours of Beceite, in the higher basin, where the pumping project was planned to be constructed. They formed a committee and launched a campaign to inform the whole basin about the details of the project and its possible negative impacts. The neighbours maintained their protest by organising a series of peaceful demonstrations and by carrying out some legal actions to halt the project. No agreement was reached in the meetings between the affected neighbours, the Central Union and the rest of the actors operating in the basin, as their interests were deeply divided. In October 1995 the EHC reinitiated the works and the neighbours from Beceite, in the higher basin, formed the Association of those Affected by the Expropriations for the Pumping Project. The association also looked for additional support among other municipalities within the basin and the Mancommunity of Municipalities of the Matarraña river basin.

By this time the conflict showed a clear confrontation between two coalitions of actors: those located in the higher basin and those located in the middle-lower basin. Whereas the irrigation communities and municipalities in the lower basin insisted on the need for water in order to save their crops, irrigation communities, neighbours and environmental groups in the higher basin were opposed to the regulation projects for its adverse environmental impacts -- it was said to alter zones protected by EU directive 78/659/CEE and the Habitats Directive -- and for being considered an inefficient project. While the two coalitions maintained polarised positions, all actors reached the agreement that the pumping project was not a final and efficient solution and that it would only serve to give some more time to find a better one. In the meantime, several actors kept searching for alternative solutions. The municipalities of the middle and lower basin led a movement pressing for the construction of the Torre del Compte dam, which would have a capacity of 29 hm^3, as a solution to guarantee water demands throughout the basin. In response to that claim, the EHC announced that the administration would study three alternatives of regulation for the Matarraña river basin: the Torre del Compte dam (capacity of 29 hm^3), the Pontet dam (capacity of 7 hm^3) and raising water from the Ebro river to the Matarraña. Meanwhile, each coalition looked for support in different arenas. Actors in the higher basin mobilised the support of some experts from the University of Zaragoza,

[1] Real Decreto-Ley 4/1995 and Real Decreto-Ley 6/1995

important regional and national environmental organisations, and some municipalities of the Aragón Pyrenees, Catalonia and Valencia. The Central Union, mostly representing the interests of the middle and lower basin, looked for the support of the regional government and started an information campaign in order to explain their scarcity problems and their reasons for supporting the pumping project. Before the end of 1997, a narrow dialogue channel between actors located in the higher and the middle-lower basins seemed to be opened. Although both coalitions had some points in common -- they all agreed on accusing the central administration and the EHC of having been unable to find an efficient solution to the scarcity problems of the basin -- still no agreement on a suitable solution was reached.

In 1998 a serious drought not only modified the definition of the water problem but also the position of some of the actors in the policy network, resulting in a new, more favourable context for negotiation. During this year the territorial conflict within the Matarraña river basin coincided with a national environmental and social movement against the construction of more dams and diversion projects promoted by the State administration. At the beginning of 1998, the creation of the Platform for the Defence of Matarraña (Pladema) enlarged the environmental coalition by the inclusion of some environmental actors operating at the regional and national levels. At the same time, the claim for a large regulation project in the middle and lower basin strengthened: several representatives of water users of the Matarraña river basin announced their support for the construction of the Torre del Compte dam, as they considered it to be the best solution for securing water supply for all users.

The beginning of the summer of 1998 saw the low basin in a critical drought. The Matarraña river basin suffered from one of the most severe droughts in recent yeas. Some municipalities suffered from drinking water restrictions and the Matarraña river had no running water in the lower basin. The Pena dam stored less than 7 hm^3[2] and the pumping project could not be used because the river had no extra flow. In response to this situation, the Central Union decided that only those fruit trees that had not been harvested would be irrigated and that illegal water uses would not be authorised. In addition, the EHC authorised a first use of the reservoir stocks of the Pena dam. Due to the persistence of the drought, a second use of 0.9 hm^3 was authorised, only the remaining 1.5 hm^3 being reserved for population supply. This was the moment, that a rupture occurred between the middle and lower basins. The municipalities of the lower basin, given the fact that they were the last

[2] The SCR estimates that approximately 6 hm3 are needed to cover the irrigation demands of one season.

users to receive water from the dam and that they had not been able to irrigate their crops yet, said they disagreed with the irrigation criteria adopted by the Central Union, withdrew the support they had previously given to the Torre del Compte project and supported the project for raising water from the Ebro river. The regional government, irrigation farmers, farm organisations and the EHC decided to co-operate to support the construction of dams and other regulation infrastructures in order to satisfy irrigation demands. At the national level, the Ministry of Environment promoted the study of the Torre del Compte dam project as well as the study of several regulation alternatives in the Matarraña river. During 1999, social conflict was still prevalent but initial signs of consensus appeared, giving hope to the possibility of reaching an agreement on a unique solution for the whole basin. In February, the higher basin decided to give its support to the project of pumping water from the Ebro river proposed by the municipalities of the lower basin. Actors in the middle basin also supported this project as an urgent measure to solve water scarcity problems. However, this did not represent a complete agreement among the three parts of the basin. City councils and representatives of the irrigation communities were reluctant because they thought its total cost would be excessive. Actors in the higher basin kept proposing other regulating alternatives such as the Pontet dam or the construction of lateral pools, and agreed on the creation of a committee to study different regulation projects in the Matarraña river basin with representatives of the three parts of the basin.

With 1999 spring and summer season's imminent water scarcity appeared again as a major problem. The Pena dam capacity was only at 7% and the water stored had to cover water demands of the 2,400 ha. of irrigated lands, the cattle raising, drinking water and illegal users. The severe drought helped keeping the dialogue open among the basin actors. In July, as the Matarraña river was dry in its middle and lower sections and the Pena Dam only stored 2 hm^3, farm production the middle and lower basin came under threat again. Maella, which was the municipality that suffered more from drought effects, had to impose daily restrictions on drinking water in order to save as much of the fruit crop as possible. As the Pena dam capacity declined to 800,000 m^3, the committee decided to study three possible emergency regulation projects: the lateral pools, pumping from the Ebro, and groundwater extraction. In August, the rest of the water stored in the Pena Dam was released in a last attempt to save the fruit trees. The persistence of the critical drought situation forced the basin actors represented on the committee to decide on one of the regulation projects studied. Their main objective was to reach an agreement between as many relevant actors as possible in order to avoid a similar situation during the next spring-summer season. The committee, now formed by the regional government, PLADEMA, the majority of the municipalities of the basin, the Central Union and a national environmental organisation, reached a consensus and chose the option of two lateral pools as the best solution. However, as the situation in the Matarraña river basin was critical again due to a new period of water scarcity, conflict between the higher and the lower basin reappeared. As an urgent response, the Ministry of the Environment announced it would finance some wells in order to extract groundwater. The municipalities of the higher basin soon announced their opposition to the project, as did the Mancommunity of Municipalities of the Matarraña river basin. In the face of this opposition, the EHC rejected the project. In April 2000, nine of the 12 municipalities in the basin, the Central Union, PLADEMA with the mediation of the *Fundacion Ecología y Desarrollo* reached agreement to promote the construction of two lateral pools. They also agreed to reject the construction of wells in the higher basin. In September 2001, the Ministry of the Environment authorised the project of the two lateral pools.

6.3.3 Property rights and governance changes

During the last decades, the extent of uses of water in the Matarraña river basin have increased somewhat. Irrigation continues to be the most important water use, although during the nineties a rivalry arose between irrigation and nature protection. This rivalry is aggravated by the fact that irrigation needs have increased due both to the change of crops in the sixties and seventies and to the decreasing availability of water due to persistent drought periods. In this respect, rivalries in relation to water use mostly take place among the same users, but are located at different territorial levels of the river basin.

The increasing demands for water and its decreasing availability are two factors that combined have led to changes both at the governance and the property rights levels. Changes at the governance level are quite internally coherent in relation to the levels of governance involved, the dynamics of the actors' networks and the problem definitions. As regards the scales of governance, during the conflicting period actors operating in the river basin established alliances with actors operating at regional, state and to a lesser extent, EU levels. For instance, actors in the higher river basin formed a coalition, gathering together academics, other municipalities and regions, as well as national environmental groups. In turn, actors operating in the middle and lower river basins obtain the support of the EHC. These actors retained different perspectives and objectives previous to the 2000 agreement. In brief, there was collusion between the traditional water supply approach and an emerging one pressing for the conservation and the rational use of the resource. While these two perspectives seem irreconcilable, drought pressure pushed the shaping of a shared perception of risk and the need to preserve water in order to secure the economic, social and environmental sustainability of the basin. The rejection of the construction of both the Torre del Compte dam and the ground water extraction wells, as well as the inclusion of environmental concerns in the definition of the lateral pools, are indicators of this. Once the 2000 agreement was reached, co-ordination among the actors operating in the river basin seemed closer than that between them and other supra-river-basin institutional actors.

The internal coherence at the property rights level seems less plausible. This is mainly due to the discord between legal aspects and current practice. For instance, the provisions introduced by the 1985 Water Act regarding concessions, records and the distribution of functional tasks are ineffectively implemented, as traditional and *de facto* use rights prevail. Apart from that, the successive regulation works, including the construction of the Pena dam and the diversion tunnel, as well as the creation of the Central Union, for instance, altered the traditional property rights system so that use rights adapt to these changes. However, adaptations of the property rights system occur reactively and in the long run.

In general terms, the relation between public governance and property right changes seems quite coherent, as changes in public governance are closely interconnected with those taking place in the property rights system. For instance, the decision to reject the Torre del Compte project, which would redistribute water uses in a traditional fashion and relax the terms for the use of water in the middle and lower basins, was abandoned and replaced by a more environmentally friendly solution agreed by all actors because there was an increasingly shared perception that the protection of the river is a key element for the future social and economic development of the basin. In this respect, the rejection of large hydraulic infrastructures and the consideration of alternative options with minor environmental, economic and social costs represents an important step forward towards the sustainability of the basin. However, while the changing of perceptions initially led by actors defending environmental concerns seems to be an important change agent, it is not the only one and two more have to be taken into account. On the one hand, the provision of scientific knowledge by actors mobilised by the environmental coalition initially formed in the higher basin helped fill knowledge gaps and legitimate environmental positions. On the other hand, the increase in the number and type of actors involved in the policy process had the impact of breaking the power concentrated in the hands of traditionally dominant actors and opened the doors to new actors and perspectives in the process.

6.4 The Mula river basin

The Mula river is a tributary to the Segura river on its right side. It is located in the Northwest area of the Murcian region, in the Southeast of Spain. The river basin has a total of 695 km^2 flowing through a length of 25 km. In order to characterise the Mula river basin, three factors must be taken into account: climate conditions, the fragmentation of the land and the traditional division between land and water property. Climate conditions in the Mula river basin are semiarid and typical of the Mediterranean basin. In general terms, both inter-annual and intra-annual climate patterns can be identified. There are five-to-eight year wet cycles followed by long dry periods. Land fragmentation is the second factor to take into account. The irrigatable lands in the Mula *huerta* cover 2,016 ha and are structured in small holdings (*minifundio*): 86% of farms, out of a total of 1,328 registered estates, each have an area less than 2 ha. The fragmentation of land into smallholdings has historically been established and is typical of the Murcian region. Land fragmentation, together with the low levels of agrarian cooperation, has largely constituted an obstacle for irrigators to transform the historical water regime and improve water distribution infrastructures. Finally, the traditional division between land ownership and water ownership has historically generated a system in which land owners have been subordinated by those few having a property title over water, the so-called Lords of Water, who have formed an institution called the *Heredamiento*. The separation between water and land property took place during the 14th century in such a way that those having a privileged position could obtain greater benefits just by acquiring water than by buying land. At least since the 16th century, the Lords of Water had property rights to about 95% of the water flowing through the *Acequia Mayor,* which is the main and most antique irrigation channel, while the remaining 5% was reserved for the neighbours. For centuries, the Lords of Water used to sell water in daily auctions and this meant they had control over the price of water and the conditions for its distribution to small land owners, subject to the historically established conditions and use rights. The 1866/1879 Water Act respected the vested water property and use rights, so that the Lords of Water kept holding the property rights to water from Las Fuentes de Mula, at the head of the river, until the late eighties. From that moment, a transformation of the water regime has taken place. To sum up, the transformation of the Mula regime into a more efficient system has traditionally been frustrated because of both the combination of land fragmentation, the separation between land and water property, and the increasingly severe situations of drought and water scarcity that have historically threatened the survival of the *huerta* (Pérez Picazo et al., 1990).

6.4.1 Property rights and *de facto* use rights

The traditional system of water property and use rights has to a great extent determined the development of the water regime in the Mula *huerta.* However, not only formal property and use rights but also traditional practices and vested rights have also played a very significant role in the Mula river basin. Regarding the former, the first Spanish Water Act, dating back to 1866, declared the hydraulic public domain of surface waters and, at the same time, established that rivers having their head in private lands or flowing through the same type of lands belonged to landowners if they decided to use those waters. Therefore, the hydraulic public domain of waters depended on the use landowners made of it. Simultaneously, the act declared the respect for the private domain of water owners and let them keep their use right and other operations of free choice, like sale or barter agreements. The same legal system was included in the 1879 Spanish Water Act and the 1889 Civil Code. Thus, in practice, water was considered as an ordinary real estate and water owners had free choice over waters. This fact allowed them to sell not only the ownership of waters but also their use right, and a way of transferring this right was by means of auctions, as in the Mula case and other villages of the Murcian Region (Héllin, 1980). In short, the 1866/1879 Water Act clearly respected these vested rights. The 29/1985 Water Act declared all continental waters as hydraulic public domain and also declared the consequent obligation of applying for an administrative authorisation or concession in case of exploiting special common uses or privative uses of water. However, this rendering water a public good had not gone as far as might have been possible as the Water Act also allowed previous owners to retain the title to those waters that were considered private waters according to the previous legislation.
In practice, informal institutions taking the form of vested rights, norms of behaviour and traditional practices have been of great importance when explaining water management at the river basin level. More particularly, uses and practices related to the water regime in the Mula *huerta* were strongly determined by the irrigation system used by farmers since the Muslims and by a water distribution system based on a vast net of irrigation channels. The core and the origin of the distribution system at the traditional *huerta* was the *Acequia Mayor*, which was the major irrigation channel, having been excavated by the Muslims in the 9th century. Over the *Acequia*

Mayor there was a complex distribution net formed by irrigation channels that distributed water to every single *huerta*. All water available was divided into 20 portions (*cuartos*), 19 of which were the private property of the Lords of Water and the portion left could be used by the municipality of Mula and its inhabitants. So the Lords of Water owned the majority of *cuartos* and could transfer the titles to water, as could be done with any other private good (Pérez Picazo et al., 1990). Farmers had to buy the *cuartos* of water in the daily auction.

6.4.2 The process of regime change

Three major periods can be identified when analysing the development of the Mula river basin regime development during the 20th century. The first one, starting at the beginning of the 20th century and ending in 1966, corresponds to the traditional system of distribution and allocation of water in the Mula river basin. The second period is defined by the transition from the traditional system to a new one, while some elements of the first period are still operating. The third one represents a clear turning point towards a new and challenging regime which aims at completing the transformation of the regime. During the first period the availability of water in the basin had been limited by the great variations in climate conditions and the property rights regime, which was controlled by the Lords of Water. They sold portions of the resource in daily auctions and had the rights of access (exclusion of non-members), withdrawal and management of the resource. During this period, the Mula *huerta* specialised in vineyards, olive and cereals because of the low irrigation demands of such crops. Irrigation was only needed once a year in winter or spring, when water could be bought in the daily auction at a lower price. The construction of La Cierva dam in 1931 and the creation of the Irrigation Community increased farmers' expectations of having increasing water and a broadening their use rights. However, severe drought periods resulted in an increasing speculation on water prices. As the property of land and water was separated, many owners sold their lands and retained title to the water. This speculative wave also affected the daily auction and traditional uses and practices. In this respect, an 'unofficial' and perverted auction following the official one in which waters left were sold to the most desperate farmers became a common practice. The competition for the acquisition of water resulted in great social conflict and in increasing rivalries among farmers.

During the post-war period, which was coincident with a not-too-dry period, the demand for more select products in the national and foreign markets induced a transformation of the traditional huerta that specialised in crops with high water demands, more particularly apricots and citrus. As a result of that, the reserves of the dam progressively decreased and scarcity problems and social conflicts reappeared once again. During this period, water from La Cierva dam, although it was administered by the Segura Hydrographic Confederation (SHC) due to its public character, was also sold at auction. In this sense, there were two types of water and two different bodies in charge of them: the private ones that were owned by the Lords of Water and were sold in daily auctions and the public ones, administered by the SHC and also sold to farmers in the auction. The main reason for this distinction was that the Heredamiento had also the right of way over the water distribution channels which implied that as all waters had to be distributed by the same net of channels, the Heredamiento determined the mechanism of distribution. From 1948, the traditional system started to come under question. Some users started claiming the autonomy of the Irrigation Community from the SHC in relation to the water from the La Cierva dam. Several proposals among farmers claiming a transformation of the traditional system of distributing water represented the first step towards an improvement of the water distribution system and use rights.

In 1966, two major changes in the traditional administrative and distribution system marked the beginning of a new phase in the development of the regime. On the one hand, the Lords of Water agreed on the transfer of water management to the Irrigation Community and, on the other hand, water auctions were abolished and the *tanda*[3] system was established. But the following years saw another drought period that resulted in scarcity problems and social conflict. The lack of supply from the Segura river concession, the high price of water and the poor efficiency of the irrigation system resulted in the rationing of the available water resources, premature crop ageing, a decrease in land productivity, and a rural exodus. As a result of the search for new water resources by the Irrigation community, in 1981 a well was opened and the concession of 4 hm^3 per year from the Tajo-Segura diversion became available. However, in 1982, another severe drought period resulted in the opening of illegal wells.

During the late eighties, and despite the improvement in the weather conditions and the continued contributions to the dam thanks to considerable rainfalls in autumn, the huerta became less and less profitable: crops were ageing rapidly, the traditional irrigation system became more inefficient, the inequalities of water distribution became wider and water prices kept increasing. Within this context, at the end of the eighties a process of profound transformation of the water regime was initiated. This process

[3] A fixed time period in which water was distributed to the *huerta*. Each *tanda* consisted of 20 days, divided into 4 periods of 5 days.

resulted in the redistribution of property rights and the adoption of a Modernisation Plan. The redistribution of property rights was directly conducted by the Irrigation Community, whereas the Modernisation Plan was elaborated by the regional Ministry of Agriculture in Murcia and the Centre for Applied Edaphology and Biology on the Segura (CEBAS), although it was based on the proposals made by the Irrigation Community.

In 1990 the president of the Irrigation Community started the process of unifying all waters by proposing that the Irrigation Community should purchase waters owned by the Heredamiento. This was possible because, in the mid-eighties, the Heredamiento had already lost its monopoly on water property rights as public waters coming both from La Cierva dam and the Tajo-Segura diversion were also available. This circumstance, together with the fact that in the mid eighties the Murcian region suffered from a severe drought period -- which meant that the Lords of Water could not get much profit from the almost nonexistent flow of La Fuente de Mula -- was a powerful reason to persuade the Lords of Water to sell their property rights to the Irrigation Community and allow the redistribution of water property rights. The Irrigation Community assigned a price to each portion of water (the cuarto) and established the terms under which it would try to purchase as many cuartos as possible with their own financial resources. Nowadays, a very reduced number of cuartos are still in the hands of the Lords of Water and the Irrigation Community keeps on negotiating with them in order to purchase the remaining ones. The process of purchasing water is expected to conclude in 2002. Once concluded, the Irrigation Community intends to make these waters public property by registering them in the Water Register of the SHC.

The Modernisation Plan was designed by the Murcian administration and implemented with the financial aid of EU funds. The plan has brought about a radical transformation of the water distribution system. To do so, the plan first updated the users' census and quantified the total water available and the monthly needs for crops. In order to optimise the distribution and use of water, the plan divided the huerta into seven irrigation sectors, each of them to receive water through eight interconnected small dams. The dams receive water from the deviation channel at the Mula river, La Cierva dam, El Pradillo and El Corral wells. Apart from that, the Modernisation Plan is innovative as the distribution net is computerised and centralised so that the Irrigation Community receives immediate information about all the operating parameters of the system and can regulate and have control over it. The water distribution system has also changed substantially, as the tanda system has been replaced by a new one based on the 'water quota', which is the minimum quantity of water per hectare available to every user at the beginning of the hydrological year. The quota is calculated at the end of each hydrological year according to the available resources and a prediction of groundwater availability. In order to rationalise the use of water, the Modernisation Plan has given users incentives to replace the traditional flooding system of irrigation with a dropping system. This has had the effect of consuming less water. In this respect, water stored in the dams that have not been used revert to the Irrigation Community. Thanks to the computerisation of the distribution and management system, users can elaborate their own Irrigation Programme. In this respect, the Modernisation Plan introduces another innovation consisting of giving each user a Water Saving Book -- which is very similar to an ordinary saving book from a bank -- in which information about water consumption is recorded. The new distribution and management system also allows the transfer of water as the Modernisation Plan has established a Water Bank operating at the huerta level. Through the Irrigation Community, users can transfer their waters from one farm, sector or owner to another. They can also transfer water to be used in future situations of water scarcity. Finally, the Irrigation Community has created a training programme for farmers which was developed at the same time as the Modernisation Plan works (Del Amor et al., 2000). In general terms, the autonomy given to users has resulted in a more efficient use of water, a dilution of social conflict and in increase of number of young farmers having a part time job at the huerta. Nowadays, the Modernisation Plan of Mula is serving as a demonstration for the transformation of other huertas at the Murcia region.

6.4.3 Property rights and governance changes

The extent of the regime in the Mula case remained relatively stable until the implementation of the Modernisation Plan. While irrigation continues to be the main use regulated in the Mula river basin, the plan has regulated other uses, mostly industry and environmental protection. Regarding industry, the plan has created the conditions for the creation of an agrarian co-operative constituted by the members of the Irrigation Community. Regarding environmental protection, the plan has elaborated an Environmental Impact Study of the Modernisation Plan and has established an ecological flow in a concrete section of the Mula river. It must be pointed out, however, that although the extent of the regime is quite complete as, gradually, all uses are regulated, in practice the ecological flow seems not to be fully respected, and thus water uses related to environmental protection and recreation might not be properly considered in the Modernisation Plan.

The internal coherence of the public governance elements has improved considerably in regard to the levels of governance, the actors and perspectives, the instruments and responsibilities for implementation. Regarding the levels of governance, a number of institutional actors operating at different political and territorial levels have entered the policy arena: the Irrigation Community, the Segura Hydrographic Confederation (national administration), the regional government, the CEBAS and the EU by means of financial aid. As the levels of governance interact, so do the different actors involved in the process. These actors have established important patterns of negotiation and collaboration in order to elaborate and implement the Modernisation Plan. They have established common objectives, designed the instruments and strategies, and assumed responsibilities in order to achieve them. By the convergence of two objectives -- improving the irrigation system at the huerta, which was a basic demand by users, and implementing an innovative irrigation system, which was a clear objective of the regional government -- these actors have designed a strategy and provided the main instruments and resources to reach such objectives. In this respect, the adjustment between objectives and strategies and resources for implementation has been very close.

In general terms, the internal coherence of property rights is rather high. Changes at the regime level include the redistribution of property rights, a process that was led by the Irrigation Community and that was in theory independent of the implementation of the modernisation plan. As it occurs in relation to the modernisation plan, changes at the property rights level are fully coherent with the objectives aimed at. Thus, the external coherence between public governance and property rights changes has considerably increased as the institutional coalition between the local and regional governments and the Irrigation Community has made both the implementation of a policy initiative and a redistribution of property and use rights among the main users possible.

To sum up, three main change agents have driven the developments in the Mula regime. First, the institutional leadership assumed both by the regional administration and the Irrigation Community. The former has concentrated political, technical and financial support in order to implement an innovative policy initiative while the latter has led the process of unification of public and private waters at the river basin. Second, the problem pressure and the consequent social demands for abolishing the privileges and monopoly of the Lords of Water has been a catalyst for the initiation of regime change. And third, the scientific knowledge provided by the CEBAS and the president of the Irrigation Community have formed a necessary input to the successful implementation of the Modernisation Plan.

6.5 The two cases in a broader context

Both the Matarraña and the Mula cases are clear examples of changes in water regimes that are attempted to improve the conditions for sustainability. However, these two cases are neither representative of water regimes in the majority of river basins nor the national regime. While the national regime has been transformed during recent years, changes have taken different directions at the national level and in the cases analysed. Focusing on variations of change is crucial in order to reach a better understanding of regime change and its implications on sustainability. To do so, six variables are taken into account in both the two cases and the national level: problem pressure, social conflict, multilevel governance, multi-actor network, institutional leadership, and scientific knowledge. It must be taken into account that the comparison is made between two different units of analysis: the river basin regime and the national regime. However, the comparison may give some insights that may be useful in order to understand trends of regime change.

Increasing problem pressure is a crucial change agent when trying to account for regime changes. In both the Matarraña and the Mula river cases, during the nineties there was an increasing perception of the risk caused by extreme drought situations, which was directly perceived as threatening the maintenance of social, economic and environmental cohesion. At the national level, there was an increasing awareness of the 'Spanish water problem' and a widely share perception of the urgent need to tackle water scarcity. Beyond this common perception, there was no shared view on the approaches and strategies proposed in order to deal with the scarcity issue. At the national level, approaches to tackle the water issue are divided into those proposing acting on the supply side and those proposing acting on the demand side. While these two opposed visions are present in the two cases analysed at the early stages of the process of regime change, actors' preferences operating in both river basins gradually converge into a shared view, highlighting the need to protect the resource.

Both in the two cases and at the national level, the multilevel nature of governance became increasingly intense. In both the Matarraña and Mula processes, new actors entering the policy arena lead to a break in the concentration of resources and to a more balanced distribution of resources.

In both cases, the river basin actors increasingly interacted with those located at the regional and State levels of governance. In addition, the EU level is indirectly present in both cases: in the Matarraña case by means of the protests of environmental groups, and in the Mula case as the EU co-financed the Modernisation Plan. The increasing presence of many actors in both processes, confirm a relatively balanced distribution of resources among actors. At the national level, in turn, the distribution of resources among actors is quite unbalanced as the national government, holding an overall majority in Parliament and having the support of some regional governments, has concentrated most of the political and strategic resources on defining a solution which does not reflect a wide social and political consensus: the National Hydrological Plan. This plan includes the construction of large regulation projects transferring water from the Ebro to the Segura river basin. Here, then, there is a differentiation among general trends in the two cases analysed and the national regime. A second differentiation relates to institutional leadership. While at the national level institutional leadership means the dominance of national government in the definition of policy priorities, in the Matarraña case institutional leadership means that the Central Union acts as an arena of negotiation and in the Mula case the coalition formed by the regional government and the Irrigation Community led to the modernisation plan and the redistribution of property rights. In these two cases, sustainability concerns enter the negotiation arena: from the bottom-up in the Matarraña case and from the top-down in the Mula case. Finally, sustainability concerns not only have a political but also a technical dimension. In both the Matarraña and Mula cases, a wide range of external and internal actors (academics, environmentalists, experts) provide scientific knowledge to the process, whereas only partial scientific knowledge is incorporated in the definition of the national policy and regime.

6.6 Conclusions

The Matarraña and the Mula river cases constitute two experiences of water regime change in contexts of scarcity. As in many of the Spanish river basins, water scarcity has become a dramatic problem and has been a source of intense social controversy. Moreover, the water issue has gone high up the national agenda. However, while at the national and, in some cases, at the regional level a traditional supply-oriented approach to the problem of water scarcity has prevailed, the Matarraña and the Mula cases have undertaken regime changes consisting of replacing this traditional approach by a demand-oriented one. As a result, the conditions for the improvement of sustainability seem to have improved in both river basins. However, the two cases have followed different strategies leading to water regime change. In the Matarraña case, the demands for a more sustainable use of water come from the bottom-up and emerge in the higher basin, having gradually opened the doors to a multi-actor negotiation pattern and a redefinition of policy preferences. In the Mula case, the regime change comes from the top-down and is led by the regional government and Irrigation Community. They effected a profound transformation of both the property rights system and public governance. In short, both cases follow very different paths to regime change. Beyond process differences, the combination of a series of factors in both cases are crucial in the understanding of regime changes. These factors include the increasing shared perception of the risk caused by extreme drought situations, the social conflict due to the collision of interests among the actors operating in the river basin, the multilevel nature of governance, the entrance of new actors into the policy arena resulting in a more balanced distribution of resources, the institutional leadership, and finally the provision of scientific knowledge by actors both external and internal to the policy process.

REFERENCES

Arrojo, P.; Gracia, J.J.; Martinez Gil, F,J.; Rubio del Val, C. (1997) *El bombeo del Matarraña en Beceite: de la ineficiencia al autoritarismo hidrológico.* Nueva Cultura del Agua. Serie Informes. Bakeaz, Bilbao.

Arrojo, P. (1998) *Puntos básicos para que los pueblos de la cuenca alta del Matarraña presten su apoyo para conseguir soluciones de futuro a la problemática de los regadíos de la cuenca baja. Universidad de Zaragoza* (Unpublished paper).

Confederación Hidrográfica del Ebro (1995) *Plan Hidrológico del Ebro.*

Confederación Hidrográfica del Ebro (1996) Aporte de recursosal embalse de Pena. Caudales medios disponibles aguas debajo de la confluencia de los ríos Matarraña y Ulldemó.

Del Amor, F., Gomez, J. and Sanchez Toribio, M. I. (2000) *Modernización de los Regadíos Tradicionales de Mula.* Murcia: Cajamurcia.

Gracia, J.J. (1999) *Estudio de aportaciones y necesidades de la cuenca del río Matarraña.*

Hellin, R. (1980) *Les huertas de Murcie*, Aix-en-Provence: EDISUD/Mondes mediterranéens.

Perez Diaz, Mezo and Álvarez, Miranda (1996) "*Política y economía del agua en España*", Madrid: Círculo de Empresarios.

Moragreg, A.A. (2000) *La vegetación en el Matarraña* (Unpublished paper).

Moreu Ballonga, J.L. (1996) Aguas públicas y aguas privadas. Barcelona: Ed. Bosch.

Nadal, E. (1984) "Aprovechamiento actual y proyectos futuros en las subcuencas hidrográficas aragonesas". Seminario del agua en Aragón. Universidad de Zaragoza, Zaragoza.

Perez Picazo, M.T and Lemeunier, G. (1990) "Los regadíos murcianos del feudalismo al capitalismo" en *Agua y modo de producción.* Barcelona: Ed. Crítica.

Riuz-Olmo J.; Lopez-Martin J. (2000) "La nutria en la cuenca del río Matarraña". Revista QUERCUS, núm 167. Enero.

Sanz, M.A., Celma F.J. (1998) *Memoria, identidad y conflicto: El agua en Valderrobes.* Facultad de Humanidades y Ciencias Sociales. Universidad de Zaragoza (Unpublished)

Sostoa, A. (1996*) Breve informe sobre las características e importancia biológicas del río Matarraña.* Facultad de Biología. Universidad de Barcelona (Unpublished).

Websites: www.mma.es
www.chebro.es
www.aragob.es
www.par.cebas.csic.es

Chapter 7

Competing Integration Principles in a Decentralising State

The Chiese and the Marecchia in Italy

Bruno Dente and Alessandra Goria
Instituto per la Ricerca Sociale (Milan-Italy)

7.1 Introduction

The attempt to create a coherent, integrated water policy in Italy has been a long and complex process, still short of being entirely successful.

There are several reasons for this, the most apparent one probably being the well known dualistic character of the Italian economy that has is roots not only in history but also in geography and climate, and therefore in the different nature of the water problems.

While the north of Italy has a relative abundance of water, the south shares with the rest of the Mediterranean area a chronic shortage, reinforced by the recent changes in the climate.

The relative abundance of water, however, has brought about the establishment in the north not only of an agriculture that consumes much more water, but also the use of the water for industrial production and hydroelectric production. This in turn has generated a situation in which quality problems are very different from one region to the other, in which there is a very different need for dams and works, in which it is very difficult to fix the 'right' price for water and in which the institutional structure in charge of coping with the problems is, to say the least, confused, ridden with internal conflicts and, as a whole, quite inefficient.

This explains why the legal regime was, until recent times, totally inadequate, as demonstrated by the fact that the basic law had no substantial

Hans Bressers and Stefan Kuks (eds.), Integrated governance and water basin management. Conditions for regime change towards sustainability, 189-212.

amendments since 1933 until the end of the '80s. This explains why two very important pieces of relevant legislation -- the first regarding water quality and the second concerning the reorganisation of the water industry -- were in fact the product of a parliamentary initiative and not -- as it is usual in Italian policies -- the result of a Governmental (that is administrative) initiative. And this explains in turn why their implementation has been clearly inadequate, and extremely slow.

However, partly because of the growing awareness of the increasing severity of the problems generated by an excessive use of the resource, partly under the pressure coming from the EU, and partly because the changing structure of the Italian government (towards decentralisation and eventually federalism), during the last 15 years efforts at pushing towards a more integrated regime have come into being.

Most of the elements characterised above are certainly shared by other countries. Even when there is a strong tradition of coping with water problems in territories that are more homogeneous from the geographical and/or economic point of view, a certain degree of fragmentation has to be expected, as well as some more or less coherent efforts to overcome it in order to bring about an integrated regime. What seems specific to the Italian situation, therefore, is neither the fragmented nature of the water regime, nor the attempts to overcome it, but rather the fact that different competing principles of integration have emerged from the legislative evolution, with different definitions of the problem, different constellations of actors, and finally, different implementation problems.

The aim of this chapter is not to explain why this is so, but rather to assess in a comparison between two cases in which the two already established forms of integration seem to be at work (the third is too recent for an adequate assessment), what factors are able to explain their relative success in terms of integration, and eventually sustainability in the use of the resource, in order to generate hypotheses, to be tested in other national or international situations, about the ways in which the generally shared aim can be reached of achieving an integrated regime, in which a plurality of different uses and users are taken into account, there is coherence between the distribution of property and use rights and water consumption is sustainable.

In order to do so the next section summarises the different types of integration that can be found in the Italian legislation, outlining the main public and private actors involved in each of them, thus justifying the choice of the cases that will be discussed in the following two sections. Section 5 then attempts a comparison, while the conclusion tries to draw some more general lessons from the research carried out.

7.2 Too many competing integration principles?

In order to give some factual basis to the claim put forward in the introduction about the huge regional variations in water problems that account for the difficulties in evolving a clear and unified water policy in Italy, let us just consider the following data:

1. from the quantitative point of view the northern regions (the Po Valley and the North east) while accounting for the 38% of the total rainfall collect the 64% of the total water available due to the large storage capacity of Alpine glaciers and lakes; furthermore, this availability is well distributed during the year while in the rest of the country between 60% and 90% of river outflows are concentrated in winter and spring; the result is that Northern regions account for the 67% of water consumption;
2. as far as the uses of water are concerned agriculture accounts for 50% of total consumption, the share being somewhat higher in the southern regions due to their much more limited supply; industrial uses on the other hand are concentrated in the north, hydropower in the North-East, and civil consumption has the largest share in the Central Regions;
3. from the qualitative point of view again, the situation is differentiated. High pollution in the North and Centre is mostly due to the industrial and agricultural activities, while in important coastal areas of the South the main problem is salt intrusion;
4. the structure of the water supply is differentiated, too: in the northern Regions is highly segmented, being basically operated at the municipal or inter-municipal level, for drinking water directly by the local authorities, for agricultural uses by farmers' associations and for industrial uses by the individual factory from the water table or the surface water. In the South the picture is totally different and the various uses are extremely interdependent since they rely on the same infrastructures -- large dams and transfer schemes -- operated by State-owned organisations;
5. finally, wide variations across the country can also be found in the price of water; despite the fact that in most recent years the increases in the water price have, on average, brought its level in line with the European standards, this holds true mostly for drinking water and in any case this average conceals huge variations from one area to the other.

It is therefore not surprising that the evolution of a coherent legislation has been slow and sometimes contradictory.

As a matter of fact, the main principle -- stating the need for public permits for water uses, both from underground and surface water -- dates back to the unification of the country (1865) and was organically revised in 1933. The Water Code of 1933 addressed in a very general way the main uses of water for production (agriculture, drainage, industry and power generation) and for consumption (both human and cattle). What was not considered by the legislation was on the one hand the uses for recreation and tourism, and on the other hand the environmental protection uses. The main actors involved were the Ministry of Public Works, the municipalities and the Consortia for Land Reclamation, i.e. the semi-public structures managed by and large by the local farmers' associations. The content of the policy was at the same time regulative (distributing the competencies to issue the permits and to regulate the waters) and distributive since it administered the large amount of funding involved in the construction and in the operation of the hydraulic works aimed at flood protection, transport of waters for civil and agricultural uses, and, more recently, depuration. Under this legislation there was, in theory, some room for private property rights, but, as a matter of fact, probably due to the Italian legal tradition, never very strong in the protection of the individual rights vis-à-vis those of the State, this was never a very important issue, in general acknowledging the dominance of the public interest, and public policy, over private claims.

Nothing important happened in the field of water policy since 1933 (and, some may argue, 1865) until the second half of the seventies, when a new piece of legislation (L. 319/1976, the so called Legge Merli) was enacted. This law dealt essentially with water quality, defining maximum emission standards, thus paving the way to the creation of a network of depuration plants both for civil and for industrial uses. Environmental considerations were at least included in water policies and this happened largely because of external pressure -- the impact of the EU -- and after the creation of the Regions. The latter point is very important, as there was not an effective national administrative infrastructure in charge of the implementation (as witnessed by the fact that the law itself was the product of an autonomous parliamentary initiative and that the Ministry of Health to which the competencies were attributed was quite reluctant and in any case ill equipped to deal with them). This meant that, in the void at the national level, the main regulatory authorities were the Regions and the local Health Units integrated at the regional level. Only after 1986 -- with the birth of the Ministry for the Environment and the subsequent creation of the network of the National and the Regional Environmental Agencies -- has there been an attempt to create a national dimension in water quality policy, but, by then, the effort was severely limited by the fact that the EU on the one hand and the Regions, on the other, had already occupied most of the policy space.

The first, and very ambitious, attempt at integration is represented by the Law 183/1989 creating the Water Basin Authorities. Here, for the first time, the effort was to take into consideration both the quantitative and the qualitative dimensions, and most of the potentially rival uses, in the direction of the integrated management of the resource at the level of the river basin. The law in fact created six National Basins Authorities to manage the most important Italian rivers, jointly directed by representatives of the national departments of Public Works and Environment as well as by representatives of the interested Regions. Furthermore 18 Inter-regional Basin Authorities, jointly managed by the interested Regions, were created, while the remaining bodies of water were entrusted to smaller Authorities under the direct control of the Region concerned. The most important task of the Authorities at all levels is to approve and implement a Basin Plan in which the compatible uses are spelled out, the priorities in water policies fixed and the monies available at the national and regional level for flood protection, hydraulic works, water transport and depuration allocated. If one reads the first articles of Law 183, one cannot avoid the impression that the Basin Plan is nothing less than a very comprehensive and very powerful territorial plan that defines all potential uses and severely constrains most urban, industrial and agricultural policies. In fact this is not totally true as the sheer complexity of articulating such a plan makes such comprehensiveness virtually impossible for large territories like the National Basins (the Po valley produces at least one third of Italian GDP). As a matter of fact, the severity of the flood protection problems, and the fact that most of the available monies are devoted to them, defines the principal preoccupation of the Basin Authorities while the environmental protection is mostly confined to the provision that safeguards the water minimum constant vital flow (introduced for the first time in Italian legislation) and the interests of agriculture are brought into the picture by the fact that very often the hydraulic works are entrusted to the already mentioned Consortia for Land Reclamation, managed by the farmers and their associations.

However the potential for integration of the 1989 legislation cannot be easily discounted. The water basin is defined as the optimal dimension for intervention, thus overcoming the problems created by administrative boundaries. The Authorities have both legal/regulatory and financial resources that can be mobilised simultaneously to cope with the problems. The Authorities improve coordination both at the centre between the Ministries of Public Works and Environment and in the centre-periphery dimension creating a tighter network of actors and increasing the potential for joint action and conflict resolution. Last but not least, the creation of technical Secretariats for the National Basin Authorities (and to a lesser extent for the Inter-regional Authorities as well) improves knowledge of the hydrological system, increasing the information resources available to all the actors.

Despite the ambitions stated in the objectives of the law, however, this attempt at integration falls short of giving adequate relevance to the problems involved in water distribution and purification, including the question of water prices and the more general issue of what we can call the 'water industry'. This is the reason why a few years after the approval of Law 183/1989 a new fundamental piece of legislation was approved -- once again a parliamentary initiative, lacking very strong support within the national administration. Law 36/1994, the so called Legge Galli, deals with water services and their management. Its main features are:

- the law states the public ownership of all water resources and sets a hierarchy between various uses of water, giving priority to human consumption;
- the law previsions the creation of an Integrated Water Service which should provide extraction, adduction, sewerage and purification;
- the law requires the definition of the optimal territorial area for the management of the Integrated Water Service; this ATO (Area Territoriale Ottimale = optimal territorial area) should be defined by the Regions on the basis of hydrographic as well as political/administrative criteria;
- in each ATO a managing body must be created, representative of the local authorities included in the area, which will be responsible for the direct provision or the contracting-out of water services;
- the re-organisation of water services within the ATO implies a revision of water tariffs, which should be unique for each ATO and should guarantee 'full cost' recovery, i.e. operational as well as investment cost recovery.

Here again integration seems to be the key word, but with a very different meaning. The main objective is to overcome the fragmentation of the water supply structure that is the dominant feature of the water industry in the most developed part of the country. This in turn implies bringing the local authorities and mostly the municipalities into the picture, which very often directly manage the aqueducts, the sewers and the depuration plants. The full implementation of the law will on the one hand force inter-municipal coordination and, on the other hand, provide the basis for an opening up of the market for water services under the compulsory open tendering required by European legislation. This in turn should bring about a more 'rational' price of water, that will in all likelihood translate into a relevant increase of the tariff. Given these features, therefore, and recalling the lack of a strong commitment within the national administration, it is hardly surprising that the implementation of the Galli law has been extremely slow and that, by the turn of the millennium, very few ATOs were in operation.

The picture would not be complete unless we recall a final piece of legislation, that is D.lgs. 152/1999, the stated aim of which is nothing less than the integration of environmental, health and economic considerations towards a global policy of water resource management. The opportunity for this law, the first major one pushed through by the Ministry for the Environment, is the need to at last implement the 1991 EU directives concerning the treatment of water effluents and the protection of water against nitrates used in agriculture. The law introduces for the first time the concept of quality objective of a water body, in line with the orientation of the future EU directive on water. In order to do so the law entrusts the Regions with the task to approve for each water body a 'protection plan' in which a 'quality objective' should be defined, on the basis of which the admissible charges, compatible with the self-purifying capacity of the body itself, will be fixed, in order to provide the baseline against which discharge limits will be specified.

In this sense the law 152/1999 overcomes the limitations of the old Merli Law and paves the way to a full consideration of quality objectives in future integrated water regimes. Given the recent character of this innovation however, no protection plans have been so far approved, let alone implemented, and it is far too early to assess the effectiveness of this new legislation. In particular it will be interesting to understand how the principle of equality will interact with the possibility that the discharge limits will be very different from one water body to the other. But this is only a reminder that property rights are an integral part of the resource regime.

Summing up the evolution of the legislation seems to point, as far as integration is concerned, in three different directions:

1. integration at the scale of the river basin is represented, mostly concerned with the quantitative dimension, and managed by a network led by Basin Authorities, a body in which the interests of Public Works, and therefore flood protection and irrigation, to a lesser extent Environment, and with a crucial role Regional Governments;
2. integration at the scale of the optimal area for water supply and depuration, mostly concerned with the establishment of the integrated water service, and therefore with water as a commodity, and managed by a network in which the Regional Governments and the Local Authorities play a major role;
3. integration at the scale of the water body, mostly concerned with the qualitative dimension, and managed by a network in which the Regional Governments and the Environmental Administration (the National Department, the National and the Regional Agencies) seem to be the key actors.

While it is still too early to understand what, if any, impact the last piece of legislation will have, it is possible to assess the first two attempts at integration. The following sections will try to go in this direction.

7.3 The Idro lake and the Chiese river basin

The first case selected aims to illustrate the first type of integration, the one based on the river basin scale and managed by the Basin Authority.

The Chiese river generates from the Adamello glacier and flows down the Daone and Giudicaria valleys, creates the Idro Lake at an altitude of 369 m. a.s.l., and then flows down the Sabbia valley joining the river Oglio 148 km from its spring. The whole basin includes two regions (Trentino and Lombardia) and three provinces (Trento, Brescia and Mantua) covering an area of 934 km^2.

The Idro Lake is a natural basin, artificially regulated since 1922. The average surface of the lake is 11 km^2, its carrying capacity is about 75,5 million m^3, while the shoreline develops along 24 km. The reason for the artificial regulation is the existence of a large hydroelectric plant downstream of the lake (at Vobarno) and of a network of irrigation canals servicing agriculture.

During the sixties, furthermore, two dams were created upstream of the Idro Lake by ENEL, the national electric company, with a total carrying capacity of more that 70 Mm^3. Therefore both the inflow and in the outflow of the lake basin is regulated, in the former case by the use of the Bizzina and Boazzo dams, and in the latter case through different hydraulic works, more precisely:

- the dam closing the lake, with a maximum outflow capacity of 220 m^3/sec above the threshold of 367 m. a.s.l.;
- the Enel Gallery, with a maximum outflow capacity of 32 m^3/sec above the threshold of 360,5 m. a.s.l.; the gallery releases the water in the Chiese River at Vobarno;
- the Farmers Gallery, with a maximum outflow capacity of 100 m^3/sec above the threshold of 360 m. a.s.l.; the gallery aims at preventing floods and releases the water in the river just after the dam.

Figure 7.1 provides a schematic representation of the whole artificial hydraulic system.

It is worth noting that the quality of the water is generally good, and this is one of the reasons why during the seventies and the eighties an important tourism industry developed along the shores of the Idro Lake, mostly catering to north-European campers.

Figure 7.1: *A scheme of the Idro lake and Chiese river water basin*

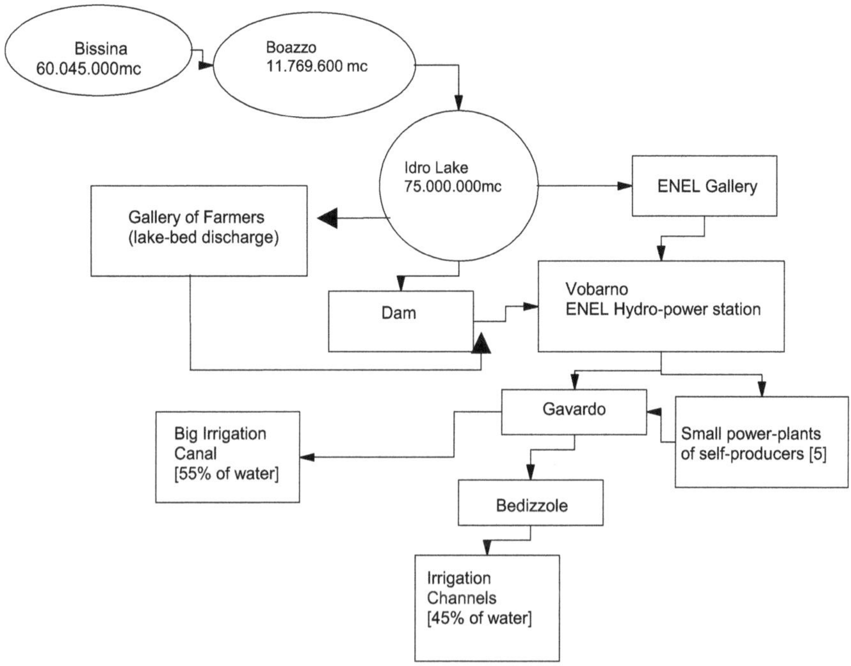

As far as the use rights and the regulatory powers were concerned it should be noted that the historical situation was the following:

- ENEL had a permanent concession to the use of the Bissina and Boazzo basins, with the obligation of releasing water in case of drought;
- the regulation of the Idro Lake through the dam was entrusted to the SLI, a limited company owned by the land reclamation consortia and in which ENEL had a share of 5%;
- the concessions for irrigation were shared by four consortia in the Brescia Province (that in 1990 merged in the Consorzio del Medio Chiese) and the consortium operating in the Mantua Province (Consorzio Alto Mantovano).

The traditional rivalries were of two types:

- the rivalry between ENEL and the farmers, due to the fact that they use the waters in different periods because of the different seasonal needs of agriculture and energy demand; this potential conflict was mitigated by the availability of the so called 'newest waters' included in the Bissina and Boazzo basins that made it possible to accommodate the normal needs of both parties;

- the rivalry between the Brescia and the Mantua farmers that was settled by the so called 'Chiese Peace' in 1953 in which was stated that of the total 27 m^3/sec of the permissible outflow, 3.7 m^3/sec were reserved for the Mantua farmers; it must be noted, furthermore, that the actual outflow was greater as *de facto* an additional 4.7 m^3/sec were available due to the use of the 'newest waters'.

But what is most striking is the fact that general rule for the regulation of the Idro Lake, the one the SLI was constrained by, defined in 1932, fixed the maximum water storage limit at 370 m. a.s.l. and the minimum at 363 m. a.s.l., thus allowing a variation of 7 meters, this created a variation in the shore line to a maximum of 50 linear meters, with obvious damages to the waterfront population, and to the tourist industry, because of the decrease in the amenity value, and the rise of hygienic and environmental problems. If one adds that the regulation did not specify the maximum allowable speed of the lake level variation, and that no provision at all was made for guaranteeing a water minimum constant vital flow (m.c.v.f.) in the downstream section of the Chiese River, there is enough to realise that there was a great potential for conflict and the emergence of further rivalries -- not as easily accommodated as the traditional ones.

The situation described above can be considered the first phase of the water regime brought about by the artificial regulation of the basin. In the general terms of the Euwareness framework it can be characterised as a very incomplete regime, as important uses and users were not involved in the regulation (waterfront municipalities and environmental issues), with an asymmetric power distribution in which the farmers played a major role and ENEL acted literally as a 'gatekeeper' with strong bargaining power.

This phase, however came to an abrupt end in the years between 1987 and 1990 when, in the first place, the concessions to the Consortia and the authorisation to the SLI for the regulation of the lake expire, and, in the second place, law 183/1989 was enacted and the Po River Basin Authority (Po AdB), of which the Chiese is a sub-basin, was created.

The first effect of those events was the emergence of new issues not represented in the previous regime. In February 1992 an environmentalist group was created (Associazione Vita Chiese) and its first request was the establishment of an adequate m.c.v.f. in the downstream section of the river. In September 1992 the lakefront municipalities, through the Mountain Community, asked for authorisation to regulate the lake -- thus taking the place of SLI -- with the stated aim of reducing the variation of water levels, and thus reducing the damage incurred by the local population and the tourist industry.

The second effect was the claim of the newly formed Po AdB to play a role in finding a solution, asking on the one hand the Ministry of Public Works to postpone the renewal of the expired concessions and authorisations until after a monitoring of the hydrological system had been completed and a m.c.v.f. for the river defined.

The following year saw the opening of negotiations between all interested parties. These meetings took place in Rome in the Ministry of Public Works and involved not only the actors representing the rival uses (the Land Reclamation Consortia interested in irrigation, the Associazione Vita Chiese asking for environmental protection, ENEL trying to protect the means of electricity production and the lakefront municipalities interested in protecting their population and the tourist industry) but also the main regional and local governments involved (the Lombardy Regional government, the Trento autonomous province, the Brescia province), plus, obviously, the Po Basin Authority.

The result of this round of negotiations was the decision taken in July 1993 by the Po AdB, in which the request for further delay to the renewal of the concession was repeated, supplemented by the appointment (again by the Ministry of Public Works) of a special commissioner in charge to enforce a new regulation rule that should -- for a three year experimental period -- introduce a reduction in the variation of the lake level from the historical 7 meters to 3.25 meters as well as a m.c.v.f. of 2.2 m^3/sec in the downstream section of the Chiese River. In order to make this possible an agreement between ENEL and the Po Adb was foreseen in which the release of the already mentioned 'newest waters' was regulated according to a curve that takes into account the probability of rain. Therefore also the maximum possible water that can be extracted by the Consortia was no longer fixed in advance but depended upon the availability of water, under the condition of no decrease in the level of the lake under 366 m. a.s.l. Finally, in order to monitor the situation, an experimental committee was set up, formed of all the potentially interested administrative bodies (at the central level the Departments of Public Works, Agriculture and Environment, both the regional departments for public works and agriculture of the Lombardy Region, the field offices of the Ministry of Public Works, the Prefect of Brescia, the Trento Autonomous Province and the Province of Brescia).

The period between the adoption of this decision and the actual start of the experimentation (July 1st, 1996) was devoted to sorting out the legal problems created by the new rule. The Consortia actually went to court in order to ask for the repeal of the decision, alleging annual damages of the order of € 75 million, but their claim was refused. In any case the Ministry of Public Works delayed the appointment of the commissioner until 1995 (again in order to resolve some legal problems involved in the creation of an experimental regime), and then postponed the start of the experimentation until the machinery for automatic flow monitoring was fully operational.

The rule adopted in 1993 was again amended in April 1996 at the request of the National Dams Service (in charge of the safety regulation of the major hydraulic works) to lower the maximum level of the lake from 369.25 m. a.s.l. to 368 m. a.s.l. because of the risk of landslide and of the collapse of the Farmers' Gallery that reduced the potential outflow of the lake. Accordingly, the agreement between Po AdB and ENEL also had to be redefined, as well as, de facto, the level of the m.c.v.f., lowered from the 2.2 m^3/sec of the 1993 decision to 1.8 m^3/sec.

The start of the experimentation marks the end of the second phase, which has seen a dramatic increase in the number of actors involved. From the perspective of the regime, what is striking is the unilateral suspension of the use rights previously allocated to the Consortia (in the case of ENEL the new constraints were defined in an agreement) as a precondition for finding a new equilibrium able to accommodate the rival uses (environmental protection and tourism) that had so far been excluded. This in turn changed the nature of the interactions from a classical case of a dyadic relationship between a technical bureaucracy (the Ministry of Public Works) and economic interest groups (the farmers and ENEL), towards a more open, multilevel network in which political actors (mainly the municipalities, but also the various provincial and regional governments) and social groups (the environmentalists, having a local constituency) are also represented. This, of course, decreased the dangers of 'capture' of the regulator, partly because new knowledge is generated through the monitoring equipment and shared with the interested actors.

The third and final phase coincided with the experimental regime and was protracted from the initial three years until the end of 2001. The only institutional change was the creation of a Users Committee in which all end-users are represented (the Consortia, ENEL, as well as the lakefront municipalities and the Associazione Vita Chiese, plus the Province of Brescia) with the task of assisting the Commissioner in defining the daily admissible flows. During the critical season -- the summer, when both the needs of the farmers and of the tourist industry are more acute -- this committee was meeting every 10 days, thus ensuring the best possible fine tuning of the general rule.

What is interesting to emphasize is that the perception of all the actors is that the experimentation has been a success, so that the extension from 1999 to 2001 was a unanimous request. This does not mean that their rivalries have disappeared: the municipalities still want to further reduce the variation of the lake level, the environmentalists demand an increase in the m.c.v.f., the farmers still claim that there is not enough water for irrigation and ENEL would like more flexibility in the use of the upstream basins. But the sheer increase in the density of the network, the sharing of information previously available only to a few actors, in the presence of an institutional mediator represented by the Commissioner (strongly supported by the Po AdB) has been enough to generate what one of the actors has labelled a 'cultural change' with the recognition of the legitimacy of the uses through a process of learning.

What it is even more important is that, despite external shocks -- periods of drought that have severely affected the interested actors -- the situation of the resource is more sustainable: the variation in lake levels is less than half what it used to be and occurs more slowly, the river downstream is vital again, as shown by an increase in biodiversity, the farmers have learnt to use less water and have improved the irrigation network.

Of course all these results are reversible. For instance, while it has been generally accepted that the regulation of the lake (the concession previously entrusted to SLI) must be shared between a plurality of actors, there are attempts to perpetuate the dominant role of the farmers. Neither is the m.c.v.f. what it should be, and the move towards less water consuming crops and/or better irrigation systems is very slow. Changes in the institutional structure -- the most important being the transfer of the concessory powers from the central administration to the Lombardy Region -- may have consequences that are difficult to predict.

But, all in all, it is a success story that bodes well for the creation of more integrated and more sustainable regimes under the umbrella of the Basin Authorities, at least in so far as the issues at stake are mostly quantitative. But we shall return to an analysis of the lesson that can be drawn after a discussion of a second case, one in which a different principle of integration seems to be at work.

7.4 The Marecchia-Conca river basin

The second case study concerns the Marecchia-Conca river basin which stretches across three Regions (Emilia Romagna, Marche and Tuscany), and four provinces (Rimini, Forlì, Pesaro and Arezzo). It includes several rivers and streams, the length of which varies between 15 and 71 km, all characterised by
high seasonal variations in surface water: in the bigger river, the Marecchia, 60% of annual water outflows are concentrated in only 45 days. This means that during the summer, partly due to the high permeability of the soil in the flatlands, the river bed is almost entirely dry.

The main importance of the basin, however, lies in the underground-water. The Marecchia cone is the largest hydrological sink of the Emilia Romagna region, accounting for more than 150 Mm^3, and with a thickness of the alluvial mattress that along the coast is more than 200 m.

As shown in the Table 7.1, the main uses of the water are largest for consumption, and even the agriculture use relies more on the water table than on the river flows.

Table 7.1: *Water Consumption in the Marecchia-Conca water basin (1998)*

Area of reference	**Civil water consumption (Mm^3/y)**	**Industrial water consumption (Mm^3/y)**	**Agricultural water consumption (Mm^3/y)**		**Zoo-technical water consumption (Mm^3/y)**
Rimini	37.0	5.9	5.9	4.7 from water-bearing strata	0.35
				1.2 from river flows	
Forlì	0.2	missing data	missing data		0.2
Pesaro-Urbino	4.5	missing data	missing data		0.25
Arezzo	0.3	missing data	missing data		0.1
Total AdB	42.0	6.8	7.1	5.7 from water-bearing strata	0.9
				1.4 from river flows	

Source: Project for a plan of the Marecchia- Conca Inter-regional Water Basin, Marecchia-Conca Inter-regional Basin Authority (Division of Water Use and Protection) ARPA, 2000

Figure 7.2 summarises the water consumption according to use in the whole of the river basin.

The huge civil water consumption in the area can be explained basically by the tourist industry along the coast, which increases the average population during the summer from 200,000 people to 2.5 million, with obvious consequences as far as water withdrawals are concerned. As far as the structure of the water supply is concerned, until 1995 it was highly fragmented: several municipal agencies managed the aqueducts of the area, other consortia provided the depuration, while the sewers were in general owned and operated by the individual municipalities.

From the administrative point of view the Basin is an Inter-regional Authority, jointly managed by Emilia Romagna, Marche and Tuscany. However, it relies mostly on the resources of the first Region, due to the imbalance in the population involved.

Figure 7.2: *Water volumes by sources in the Marecchia-Conca river basin*

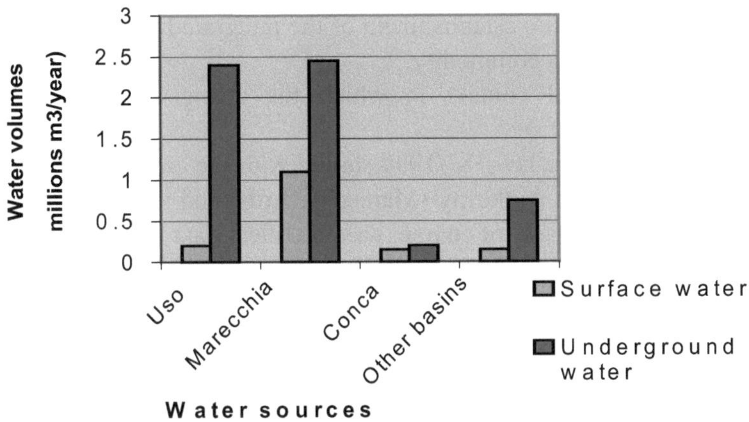

Source: Project for a plan of the Marecchia – Conca Inter-regional Water Basin, Marecchia-Conca Inter-regional Basin Authority- ARPA Emilia Romagna, June 2000

As far as the rivalries are concerned two, both relating to the underground water, are of paramount importance.

The first is quantitative in nature and concerns the impact that the withdrawals from the water table have on the subsidence. As a matter of fact it has been demonstrated that the traditionally unsustainable use has already caused major problems along the Adriatic coast (the best known example being Venice, of course). In order to prevent a further worsening of the situation large investments have been made. In particular a 300 km aqueduct bringing fresh water from the Ridracoli dam in the Appennines has been built, partly to ease the Rimini area's problems. Furthermore, a project of extending the Emilia-Romagna Canal (CER) that already exists, for using the Po water to cover the irrigation needs of the region, has been on the table for at least 20 years. As far as the Marecchia aquifer is concerned, an overall decrease of 10 Mm^3 in the last 10 years has been suggested by some official studies, but it is hard to estimate the level of withdrawals as there is no inventory of existing wells.

The second conflict is of a qualitative nature and basically concerns the level of nitrates in the water table due to the use of fertilisers. Measurements have demonstrated than the situation is already critical, but even here is difficult to pinpoint the responsibilities.

In any case, the result of the two rivalries is to put stress on the main water use, both from a quantitative and from economic point of view, i.e. the civil consumption for drinking water in the Rimini area. It is hardly surprising, therefore, that what we find in this case study is an attempt to implement the second form of integration outlined in the second section: the integration at the scale of the optimal area for water supply and depuration, mostly concerned with the establishment of the integrated water service, and therefore with water as a commodity.

This is the general context in which the changes in the national legislation take place.

The enactment of the law 183/1999 started with the creation, in 1992, of the Inter-regional Basin Authority (Marecchia AdB) and the appointment of its Secretary. The first task, of course, was to collect data and information at the scale of river Basin, and this was done immediately by mobilising the Regional Environmental Agency with a preliminary study that would eventually be transformed into the 'Project for a Plan of the Marecchia-Conca Inter-regional Water Basin', released in June 2000. In the meantime the creation of the institutional structure took an inordinate amount of time, and the Institutional Committee, that is the governing body, was to be appointed only in 1999. The single officially recorded activity was the preliminary approval in May 2001 of a plan mapping the hydraulic risk in the area and stating the priorities for investment in this sector. The reasons for this apparent lack of activity are probably on the one hand the institutional complexities of inter-regional co-operation, but also the fact that the problem pressure is relatively slight, if not in absolute terms, at least in relation to the real issue at stake, i.e. the qualitative and quantitative problems involved in the supply of drinking water to the coastal area, an issue, it must be noted, for which the Marecchia AdB has no direct competence.

Much more spectacular are the developments of the second type of integration described above, the one concerned with the creation of the integrated water cycle. In fact, the first effect of the law 36/1994 (the Galli Law), is a seemingly independent event, that is the transformation of the municipal agency in charge of the Rimini aqueduct into a limited company, owned by the 24 municipalities of the Rimini Province plus a certain number of private and public bodies. The birth, on 1st January 1995, of Amir spa is clearly linked with the idea of using the experience accumulated by the municipal agency for creating the 'integrated water cycle' manager previsioned in the Galli Law. In 1999 Amir merged with the Consortium for Water Purification of Rimini and started to enlarge both the number of municipalities served and the number of goods and services that it provides. By 2000 the number of clients had risen from 57,000 to 211,000 and Amir provided every service connected with water, from extraction to distribution, sewerage and purification (but they also produce and sell bottled water and maintain public fountains), plus a number of unrelated services (management of cemeteries, optical fibre networks). It is apparent that the aim is to create a multi-utility company to increase the capital value of the existing assets.

The same path, in a more limited way, was followed by other municipalities of the river basin among which the biggest are Riccione and Cattolica, who in September 1995 give birth to SIS spa, a company that by 2000 manages water distribution for 30,000 people as well as sewers and depuration.

The result was that when in 2000 the ATO for the Rimini province was created (one of the first in Emilia Romagna and in Italy) there were two obvious candidates for the management of the water cycle. At this point there was a rationale for a merger, that in fact took place by the end of 2001, through the creation of Adria spa, with the project to list the company on the stock exchange. After a few months, however, the perspective changed drastically as Amir and SIS were in fact merged into a new holding -- HERA spa -- collecting the utilities in all the Region, starting with the powerful SEABO, the water and energy company owned by the Bologna local governments.

In few other cases has the Galli Law been so successful in changing the structure of the water industry, overcoming the traditional fragmentation in managing the water cycle. It can be expected that this has improved the efficiency of the water schemes, modernised the network and improved the quality of depuration. But has it contributed to the solution of the existing rivalries in the direction of a more sustainable use of the resource?

The evidence collected suggests that it has not, and that, on the contrary, it has created new problems.

Starting with the second rivalry -- the one about water quality and the use of fertilisers -- there is no evidence whatsoever that anything, apart from commissioning a new study, has been done. It is obviously in Amir's interest to close down most of the 15,000 wells that are used for domestic water consumption, since more than 98% of the area is covered by aqueduct pipes. This could also have some positive impact on the underground pollution and it would decrease water consumption. But of course there would be social opposition to such a move, and the water providers seem unwilling to enter a fight they are not certain to win.

On the contrary, a link seems to exist between what happened and the worsening of the situation in the rivalry concerning the quantity of water extracted from the underground table, where the subsidence is the main environmental problem. Not only there is no evidence of a reduction in the withdrawals, the price of water has also not risen in order to incorporate the true environmental costs. This means that the water reserves are still decreasing at a rate that is probably unsustainable even in the short term, if the level of tourism is to stay constant and the climate stays dry. This is particularly striking because, in fact, there would be an alternative source of high quality water -- the Romagna aqueduct bringing it from the Ridracoli dam -- but this is not used by Amir to its full potential. In the mid nineties there was an even conflict between Amir and Romagna acque, the company owned by the local governments that owns the dam and the aqueduct, because Amir was using 5 Mm^3 less than the amount of water it was contractually bound to withdraw. Even today the water from Ridracoli covers only less then 16% of the needs, while the water extracted from the Marecchia cone accounts for 75% of the needs. The official standpoint of Amir is that there is not a quantity problem in the cone, that the official figures for the agricultural withdrawals (4 Mm^3/year) are under-evaluated as there are a lot of non-monitored wells, and that in any case the subsidence problem is due to the gas extraction from the Adriatic Sea. As a matter of fact, the problem is pretty much economic in nature: the Ridracoli water costs around € 0.45 per cubic meter against the € 0.10 for the extraction from the Marecchia water table. In a situation in which the main objective of the water industry has been to improve short-term profitability in order to convince the Municipalities owning the infrastructures to sell them at a premium, and to collect financial resources on the capital market, it is obvious that any decrease in consumption, or a supply from more expensive sources is not an option. Of course it would be possible to increase the water tariff, but this too is unlikely, as it should be done by the local governments governing the ATO, who will be blamed by the population.

The lesson that can be drawn from the relative success of the integration at the level of the ATO is that there is no evidence of a positive effect on sustainability, and that there are clues pointing in the opposite direction. The use of market forces to bring about a more sustainable use of water probably has severe limitations and can occur only if on the one hand a clear water policy is already in operation and, on the other, if the price reflects the true (marginal) value of the resource. The latter condition is always difficult to meet, even more so if political considerations play a major role (and here the weight of the local governments in the ATOs is clearly a problem) and in any case not in the first phase of the transformation in which the profitability goal can be achieved much more readily through an increase in quantity and in tapping the cheapest sources.

Again, this conclusion must be qualified. The increase in efficiency and the rise of economies of scale can actually achieve a better consumption of the resource, while the opening up of a competitive market could help to push the price of water to a level incorporating the total costs, including sustainability. But this is neither an automatic effect of integration, nor something that can take place in the void of public environmental policies.

The weakness of the integration at the level of the river basin shown above, and the still early days of the third type of integration (the one at the scale of the water body, mostly concerned with the qualitative dimension, and managed by a network in which the Regional Governments and the Environmental Administration seem to be the key actors) seem from this point of view much more relevant in explaining the unsatisfactory results of the Marecchia policy.

7.5 Regime change, integration and sustainability

The two cases are very different from each other, but they can be correctly considered good examples of an effective implementation of two different attempts at integration regarding respectively the scale of the river basin and the scale of the integrated water service. This is rather obvious in the case of the Chiese River, but the same also holds true in the case of the Marecchia if one recalls that it has been one of the first ATO to be created and that the merger of the different service providers has been nothing short of spectacular, at least in comparison with what happened, or did not happen, in the rest of the country.

In this section we shall try to characterise the two cases in the terms of the conceptual framework used in the international research, starting with the assumption that in both cases there has been a conscious attempt to bring about integration as defined in the Euwareness proposal.

If we look at the outcome of the process, there is little doubt that in the Idro/Chiese case this attempt has been successful. Not only has there been an increase in the interaction between levels of government, an increase of the uses and users brought into the governance system, but the "interrelatedness of different aspects of the problem and their dependencies" has been recognised and intensely debated in order to set common goals. There is also some evidence of a stable institutional setting in which the interaction takes place (the birth of a policy community) and, all in all, one should conclude that the actors acknowledge that integration is necessary (also) in order to prevent further deterioration of the resource. Furthermore, the knowledge of the dynamic of the water body has increased through the monitoring equipment, and is by now shared by all the relevant actors.

Not everything is perfect, and some of the results are clearly reversible, but the claim that this is a success story seems well founded.

If we ask ourselves which factors account for this change we find a couple of hypotheses in the conceptual framework that fit well with the evidence collected. Certainly the rise of the Po AdB has provided an effective institutional interface for the governance dimension of the integrative attempts and probably there has been a growing awareness of the fragility of the previous regime due to the excessive imbalance in the distribution of the benefits between the different users (very great benefits for the farmers and ENEL, virtually no benefits at all for the tourist industry, the local population and the environment). Increasingly there is an awareness of the 'jointness of the problem', even between the most reluctant users, i.e. the farmers.

Another relevant factor, again related to the very notion of a regime, is the sharp decrease in the value of the property rights component, due to the expiry of the concessions for water use and for the regulation of the lake level: the ability to exploit this window of opportunity has been crucial. Finally, a facilitating factor, in line with the general theory of public policy analysis, has been the fact that the redistribution that has taken place was not a zero-sum game, since the release of the newest water by ENEL was able to increase the total dimension of the cake to be shared, thus easing the reduction in water consumption by agricultural uses.

Having said that, the inclusion since the beginning of the environmental considerations in the governance component of the regime seems of paramount relevance in explaining the fact that sustainability (at least in the form of the minimal flow) has been included in the policy goals. This has happened because of the willingness of the Po AdB to stress this aspect and because of the existence of a grass roots organisation in the area committed to the protection of the environment, in the total absence of the environmental administration and of the national NGOs. The inclusion of the local environmental groups in the first phase of negotiations and in the Users' Committee during the experimentation phase seem to be a critical dimension in the relationship between the regime change and the increased sustainability outcome. If this element were absent, the outcome could very well have been some sort of compromise between the lakefront municipalities and the farmers, with little benefits for the environment.

The last considerations are useful in order to avoid too simplistic an interpretation of the case. The creation of the AdB -- the attempt to integrate at the level of the river basin -- has certainly been very relevant in the developments that took place in the Idro/Chiese water system. But the outcome cannot be conceptualised simply as the successful implementation of a national law. Without deliberate attempts of motivated actors going in the same direction neither the 'real' integration nor the environmental benefits would have occurred.

If we now turn our attention to the Marecchia case, the first conclusion that can be drawn is that no real progress can be detected, as far as the rivalries concerning water quality and quantity are concerned. On the contrary, there are clues going in the opposite direction. A new, and more sustainable, equilibrium could have been possible (buying more water from the Ridracoli dam and/or enlarging the CER canal), but this has also been ruled out of the new organisational setting created by the integration at the level of water cycle management.

In the policy theory underlying the Italian legislation this should not have happened. In the Marecchia case the two levels of integration -- the one at the level of the river basin and the one at the level of the scale of the optimal area for water supply and depuration -- should have reinforced each other.

As a matter of fact, it is possible to argue that the weakness of the AdB, and its ineffectiveness in dealing with rivalries related to water, was also the result of the minor attention that the actors involved in both integration processes -- the regional, provincial and local administrations -- were paying to this dimension of the problem vis-à-vis that of the question of the integrated water cycle. And there is little doubt that the economic values implied in the latter issue were far larger than the ones involved in the basin planning.

The emphasis has always been on the major economic use of the resource: providing drinking water to the tourist industry along the coast. The process of mergers and consolidation, the real aim of the privatisation of the municipal agencies and consortia, has certainly been successful in bringing about a decrease in the fragmentation of the water supply, but has also created a very powerful actor both from the economic and the political point of view, if only because the ownership of the company is still very much in the hands of the local authorities. In the terms of the conceptual framework we have adopted, this means a strong concentration of property rights and an increase in the incoherence of the regime. The local authorities find themselves in an impossible situation: they are at the same time the owners of the new company, interested in its short-term profitability, the representatives of a community formed by the clients of the company itself, obviously interested in paying as little as possible for the water, and the guardians of public interest, that is of the sustainable status of the resource. In the short term, as is to be expected, it is the latter role that suffers.

Here again a hypothesis put forward in the conceptual framework seems to be verified. When there is a great economic benefit stemming from an unsustainable resource use, and the policy field is narrow, giving rise to small and few collective action problems, a strong public governance intervention is very desirable in order to achieve a more sustainable consumption. The latter element is clearly absent in the Marecchia case, while the consolidation of the previously fragmented structure of water supply can be equated to the existence of few and small collective action problems. In other words the implementation of the Galli Law, in a context in which the environmental dimension is weak both at the institutional level and in the society, bringing about the 'integration' of the water supply structure, strengthens a set of actors whose interests, in the short term, run against the development of sustainability. From an analytical point of view the most interesting conclusion is that once again we find a situation well known to the public policy analyst, but too often forgotten in policy design, that is, that policies can fail to achieve their goals not only because the implementation structure is too complicated but also because it is too simple.

7.6 Conclusion

Should we draw the conclusion that the policy theory underlying the attempts to 'integrate' the management of the water cycle is wrong and counterproductive in terms of resource protection? This would probably be going too far. It is not necessarily so, and under different circumstances the outcome could very well be reasonably positive.

But the point is that, unless the environmental dimension is directly represented by some actor in the governance system, to expect automatic improvements in sustainability simply because, e.g., the structure of the water supply is less fragmented, seems to be overly optimistic.

An effective governance system must be based upon a large policy community, possibly reinforced by some organisational or procedural institutional arrangement, in which all the dimensions that should be integrated are represented by some specialised actors, while others play a distinctively integrative role, at least in the form of providing the necessary mediation to the interaction. Within this setting, in which different levels of government and the various sectoral issues should be equally represented, the fact that some actor has too many roles to play -- like the municipalities in the Marecchia case -- can be viewed as a major problem that will end up in an unstable, and very likely unbalanced, regime (because they will have to chose which role to play).

In the Italian case a possible conclusion that can be drawn is that, while all the attempts at integration recognizable in the evolution of the national legislation address real problems, they, paradoxically, do not seem integrated with each other, not so much because there is not a hierarchy between them, but rather because they give too many roles to play to the same actors.

From this point of view it will be interesting to observe, in a few years time, what will be the outcome of the third attempt at integration, the one at the scale of the water body, mostly concerned with the qualitative dimension, and managed by a network in which the Regional Governments and the Environmental Administration seem to be the key actors. We actually suspect that the omens are bad: it will introduce a third scale of integration (the water body that can very well be different both from the basin and from the optimal area for the management of the water cycle), will be centred on an administrative structure -- the environmental administration -- very weak from many different points of view, and it will increase the stress on the Regional governments, giving them a further new role to play.

All in all, from the point of view of institutional design, the superiority of the first attempt -- the creation of the Basin Authorities -- is apparent. This not only seems to be the only one in which different points of view are represented, and the one that has more instruments at its disposal (regulation, financial transfers, mobilisation of knowledge), but it is also the one in which the existence of permanent Secretariats creates the institutional incentives to increase multi-level and multi-sector cooperation and eventually integration. Of course -- as the Marecchia case shows -- the mere formal existence of the Authority is definitely not enough. But unless some sort of institution -- bringing together different issues and different levels -- is created, the rise of integrated regimes will be slow in coming, partial in scope, and fragile in prospects.

The very final point to be addressed is about the property rights component of the water regime. Contrary to our expectations -- and to the expectations of the legal profession that, even before 1994, was quite convinced that the public interest was paramount -- we have to conclude that property rights do matter. In the two cases we have analysed they matter in a negative way, in the sense that it is the expiry of the concessions that creates the necessary window of opportunity in the Idro/Chiese case and that the consolidation of the water supply structure operated by Amir in the Marecchia case has prevented the rise of a truly integrated regime. A very simple explanation is that because they reflect vested interests, usually with a strong economic component, they obviously make change -- and therefore change towards integration -- more difficult. As sometimes happens, the simplest explanation is probably the best, but this means that one should expect that this will occur more generally, and that, *ceteris paribus*, in order to overcome the problems created by the existence of these rights, some sort of economic compensation is probably in order.

REFERENCES

AMIR s.p.a. (2001) Studi e ricerche idrogeologiche ed idrodinamiche a supporto della gestione delle risorse idriche del 'Bacino Marecchia'.

AMIR s.p.a. (1994) *La risorsa acqua a Rimini. Un decennio di ricerche e studi.*

ARPA, Emilia Romagna (2000) *Progetto di piano del bacino interregionale Conca e Marecchia,* Autorità interregionale di bacino Marecchia-Conca.

Autorità di Bacino del Po (1994) *Stato di Avanzamento della Pianificazione di Bacino- Sintesi dei Principali Risultati.*

Barbero, G., Bertoli, L. (1998) *L'influenza del deflusso minimo vitale sulla regolazione dei grandi laghi prealpini*, Series Acqua Uomo Terra.

Bonomi, A. (ed.) (1996) *Idro e il suo Lago. Documenti e itinerari nella storia di una comunità*, Comune di Idro.

Comunità Montana Valle Sabbia (2000) *Piano di Sviluppo Socio-Economico.*

Fazioli, R., Massarutto, A. (1998) *La leva tariffaria per l'uso sostenibile dell'acqua*, Atti della Conferenza Nazionale su Energia e Ambiente, ENEA, Rome.

IRSA-CNR (1994) *Un futuro per l'acqua in Italia*, Rome.

Lugaresi, N. (1995) *Le acque pubbliche*, Giuffrè, Milano.

Malaman, R., Cima, S. (1998) *L'economia dei servizi idrici*, FrancoAngeli, Milan.

Massarutto, A. (1999) *Agriculture, Water Resources and Water Policies in Italy*, Feem working paper, Milano, n. 33.

Massarutto A., Pesaro G. (1996) 'Meccanismi istituzionali per la definizione e attuazione della politica dell'acqua nel Piano di Bacino del Po: un'analisi di casi', IEFE Notebooks, Bocconi University, Milano.

Massarutto, A., Pesaro, G. (1995) *La pianificazione di bacino come politica pubblica: il caso del Po*, Iefe Notebooks, Bocconi University, Milan.

Provincia di Rimini (1999) *Il Piano Territoriale di Coordinamento Provinciale.*

Regione Emilia-Romagna (1992) *Piano delle acque del circondario di Rimini.*

Chapter 8

Rivalry Based Communities in Europe's Water Tower

The Valmaggia and the Seetal Valley in Switzerland

Corine Mauch and Peter Knoepfel
Institut de Hautes Études en Administration Publique (Lausanne-Switzerland)

8.1 Introduction

Due to the fact that precipitation in Switzerland is approximately twice the average European value, that some six percent of Europe's total freshwater stock is stored in Swiss water bodies, and due to its geographical position in the central European Alps, Switzerland is described as the 'water tower' of Europe. Nevertheless, water management in Switzerland currently faces several challenges. The most important ones are: (1) the problem of increasing competition of (mostly heterogeneous) rival uses of water; (2) the problem of water quality (related to diffuse pollution); (3) the question of minimum residual flows (mainly in the context of hydropower production); (4) the problem of the increasing imperviousness of soils in settlement areas; and (5) the question of natural hazards relating to water and of protective measures respecting ecological needs.

The Swiss political system is characterised by direct democracy, its distinctive federalist structure and its political and societal pluralistic system. Even if over the past century, tasks have been increasingly assigned to the Confederation, the Swiss cantons still exercise a great deal of influence and power in the political arena since the implementation of most of the public policies regulated by the Confederation is assigned to the cantons, often with considerable room for manoeuvre. Thus, the main public actors in Swiss water policy are the Confederation, the cantons and the local authorities.

The emergence of regional institutional water regimes in Switzerland, their impact on the relevant actors' behaviour and on the state of the resource

Hans Bressers and Stefan Kuks (eds.), Integrated governance and water basin management. Conditions for regime change towards sustainability, 213-245.

itself, in a second phase, was investigated by means of two case studies: the Valmaggia and the Seetal valley. In the first phase, the national determinants of these regional water regimes had been analysed and, at the level of the Confederation, three main issues in water policy had been identified: 'protection against water' (i.e. flood protection), use of water for energy production, and the protection of water (cf. Mauch and Reynard, 2004). The two case studies cover all of these main issues.

The following section describes the national context of regional regime development (section 8.2). In sections 8.3 and 8.4, the two case studies are described with respect to the evolution, the triggers and the impacts of regime change at regional level. Section 8.5 then presents an analytical comparison of the two case studies. On this basis, in section 8.6, conclusions are drawn with respect to the emergence and the impact of the Swiss institutional water regimes.

8.2 Three branches of Swiss water policy: substantial integration and persisting institutional fragmentation

The Swiss regulative system (*property and use rights*) is mainly defined at three levels: the Swiss Civil Code (enacted in 1912), the Federal Constitution (Cst), and federal laws. Rights to the ownership and use of water are regulated by the two general principles of 'private property' and 'state sovereignty'. The principle of private property is defined in article 667 of the Swiss Civil Code (CC) which extends the possession of land to the areas below and above it. The principle of state sovereignty with respect to water (*Gewässerhoheit*) restricts private property by reason of the prevailing public interest. Furthermore, the Swiss Civil Code makes a distinction between public water bodies (article 664 Civil Code) and private water bodies (article 704 Civil Code). The public water bodies include surface waters (rivers, streams and lakes) as well as glaciers and firns. The cantons are responsible for the regulation of use rights over surface waters (article 664 CC and article 24bis Cst). Thus, the surface waters in all cantons are considered public property.[1] Water sources are basically considered private waters since they represent an integral part of the ground on or under which they are located. However, sources rising from a glacier, some major sources of general interest and sources at the head of a river or stream are all considered public property. Similarly, underground waters are now generally

[1] With the sole exception of the canton of Glaris where surface waters are considered private property (Leimbacher and Perler, 2000: 262).

considered as public property (Leimbacher and Perler, 2000: 260). In general, use rights to a resource under state sovereignty are assigned by means of permits (e.g. for sailing events on lakes), licences (e.g. for fishing) or concessions (e.g. for hydroelectric power production), mostly by the cantons or local authorities.

For historical reasons, Swiss water policies have mainly developed along three different branches of *water policy* during the past century (cf. Mauch et al., 2002; Reynard et al., 2000). As a first issue, after several catastrophic floods in the second half of the 19th century 'protection against water' (i.e. flood protection) was regulated at the level of the Confederation, followed by a national legislation relating to the use of water for energy production at the beginning of the 20th century responding to technical evolution. With the emergence of water quality problems in many parts of the country due to growing population density and industrialisation, mainly after the Second World War, protection of water was introduced as a third issue in water policy in the 1950s. These three branches still form the basic structure of Swiss water policy today, which is mirrored in three respective administrative branches at the level of the Confederation[2] and in the cantons as well.

The analysis of the (national) determinants for (regional) regimes resulted in the identification of a last major *phase of change* extending from 1975 to 1991 (cf. Mauch et al., 2001; Reynard et al., 2000; Mauch et al., 2000). In legislative terms, the beginning of this period is marked by the adoption in 1975 of a new article in the Federal Swiss Constitution which added a quantity dimension (mainly involving residual flows) to the existing protection of water quality. The end of the phase is formally marked by the adoption of the new Federal Law on the Protection of Waters in October 1991 which finally substantiated the principles defined in the constitutional article of 1975. However, in the case studies, this time frame is extended to the end of the 20th century as the (national) determinants of change are expected to require some time to become operational at local level. After the phase of transition from 1975 to1991, the range of goods and services regulated at the national level reached a very high level as, with the exception of those involving the preservation of water quantities, which were exempt from the intervention of public policies up to 1991, all of the goods and services provided by the resource, which are known to date, were

[2] To date, the respective administrative branches mainly responsible at federal level are: The Federal Office for Water and Geology (*BWG*, located in Biel) for 'protection against water' (i.e. flood protection), the Federal Office for Energy (*BFE*) for the use of water for energy production, and the Water Protection and Fisheries Division of the Swiss Agency for the Environment, Forests and Landscape (*BUWAL*) for protection of water (cf. Mauch et al., 2000: 2).

affected. Through the introduction of the Federal Law on the Protection of Waters, the gap which existed during the previous phase was finally bridged in the 1990s as actors drawing water for quantitative uses (e.g. irrigation, hydroelectric power) and farming activities as source for diffuse pollution were now also considered as target-groups of the Swiss water policy.

There was no formal modification of ownership rights as defined in the Swiss Civil Code during recent decades. The only change in the *regulatory system* during this phase took place at the level of the organisation of use rights. Following the introduction of a major new restriction on users of the water resource under article 24quater of the Constitution in 1953 on the protection of water bodies against pollution, the protection of water bodies applied to all kinds of water, irrespective of their property status -- private or public. The revision of Article 24bis of the Constitution in 1975 added new restrictions to the use of water, particularly with respect to hydroelectric power, by instituting the principle of the quantitative protection of the hydrological system. In 1991, the Federal Law on the Protection of Waters substantiated this by imposing an obligation to maintain suitable residual flows for water bodies. The 1997 revision of the law saw the formal introduction of the 'polluter-pays' principle into Swiss water protection policy.

The development of the *objectives* in *water policy design* reflects the evolution of its three main branches (i.e. protection against water, water use, water protection). The adoption of Article 24 of the Constitution in 1975 marked an important turning point, since for the first time it enshrined the principle of the 'unity of water management' and thus dealt with the three branches of water policy simultaneously. Formally, the main means of using the water resource were connected to each other by means of placing restrictions on one use 'in the interest of' other uses. For example, the drawing of large quantities of the resource for the purpose of hydroelectric power production was newly restricted under a protection objective relating to the hydrological cycle[3] and nature conservation considerations.

With respect to the *instruments* existing at the national level during this period of change, financial means and subsidies were generally only applied for protective objectives.[4] In contrast to the aim of protection, in general the economic use of water was regulated by means of property rights arrangements (e.g. licenses, permits) which guaranteed the role of the state in the exploitation of the resource (state sovereignty, public waters). There is, however, one major exception to this rule: the economic aims of

3 "Haushälterische Bewirtschaftung der Wasservorkommen", i.e. economic management of water resources.

4 Protection 'against' water and later also 'of water', first for wastewater treatment plants and later also subsidies for other measures.

agriculture received indirect financial support by way of subsidies for drainage and other improvements throughout most of the reference period up to the early 1990s. It was not until the later decades of the century that further restrictions on specific economic uses were introduced (hydroelectric power and agriculture, e.g. water protection zones, restrictions on fertiliser use).

The *target groups* in the policy design have evolved in accordance with the changes in the causal hypothesis. The prevailing hypothesis for this period ("If water is protected globally, its sustainable use is guaranteed"; cf. Reynard et al., 2001) only took effect very gradually during the later decades of the period. In the 1970s, actors producing 'concentrated' pollutant loads (industry, households and public bodies) represented the target groups of the water protection policy whereas, from the early 1990s, all users having an impact on water quality, the preservation of sufficient water quantities and the hydrological system were regarded as the target groups of the new Law on the Protection of Waters. Thus, the scope of the target groups was extended and diversified and agriculture also became an important target group.

All in all, it is possible to observe a distinct reinforcement of efforts to establish co-ordination between the different policy fields, particularly from the early 1990s. Increasingly, the articles in specific laws relating to water and affecting other policy fields were simultaneously introduced into the regulations governing the related policy field.[5] By this means, far-reaching co-ordination efforts were consciously implemented and institutionalised. These inter-policy co-ordination measures can be interpreted as institutional attempts to resolve rivalries between the different uses of water. A cursory analysis of the political decision-making processes leading to the new water regulations at national level during the last decade reveals the following rivalry issues as the most contentious (cf. Mauch et al., 2001): hydroelectric production vs. nature conservation and landscape protection (mainly residual flows); hydroelectric production vs. fishery and tourism (residual flows); nature conservation and landscape protection and tourism vs. flood prevention; drinking water and nature protection vs. agriculture (diffuse pollution); flood prevention vs. agriculture.

Even if Switzerland has not yet become a member of the European Union, as a result of close economic and trade relations with the EU, Switzerland is in a way 'silently' and gradually adapting to European standards and directives. This also holds true for water-related issues and the

5 E.g. Federal Law on Agriculture, Federal Law on Fishery, Federal Law on Nature Conservation, Federal Law on the Policing of Dams, Federal Law on the Protection of the Environment, Federal Law on the Use of Hydroelectric Power .

European Water Framework Directive is taken as an important guideline for further developments in water policy.

The following sections describe two Swiss case studies investigating water regime change at a regional level and their impact on the sustainability of the resource uses. For each sub-case, its location and context and the existing water uses are described, followed by an overview of the main use rivalries in the region and the respective elements of the institutional regime. On this basis, the change of the regional regime is analysed with respect to its impact on the sustainability of the water uses.

8.3 Inventing quantitative water protection: The Valmaggia valley[6]

8.3.1 Water uses in a southern alpine valley

The Valmaggia valley lies in the canton of Ticino in the southern part of Switzerland. It incorporates 22 municipalities. The upper Valmaggia consists of three tributary river valleys and the lower part ends in a vast delta plunging into Lake Maggiore at the mouth of the Maggia river (see Figure 8.1). In this delta, two towns (Locarno and Ascona) frame the river, as it is channelled into its final stretch. Economically, the valley is exploited by means of tourism; stone, which is quarried to make construction materials; and, finally, water, which is used to produce energy. There are no significant municipalities in the Valmaggia itself as the population density is very low (fewer than 20 inhabitants per km^2). Finally, agriculture plays only a very minor part in the economy, as the morphology of the terrain, hemmed in by steep hills, is inhospitable to this activity.

The water basin of the Maggia covers an area of approximately 930 km^2. The course of the river runs along about 50 kilometres from the Naret lakes, at an altitude of 2,240 metres, to Lake Maggiore, at 193 metres. Its main tributaries are the Lavizzara, the Bavona and the Rovana in the upper part of the water basin. Two other tributaries, the Melezza and the Isorno, flow into its waters at its final stretch. Its last deltaic section is heavily dyked to protect the inhabitants and the infrastructure. The rates of flow are strictly linked to the pluviometry. They reach their peak in April and May, accentuated by the melting snow; values are also very high in October. In

[6] This section relies on the Euwareness report 'First case study for Switzerland: The Valmaggia' by A. Thorens, delivered to the EU in Spring 2002 (cf. also Thorens, 2002).

terms of quality, we should stress that the waters of the Maggia count amongst the cleanest in the canton.

Figure 8.1*: Overview of the Valmaggia water basin with its main elements relating to the water resource, its various uses, and its institutional regime.*

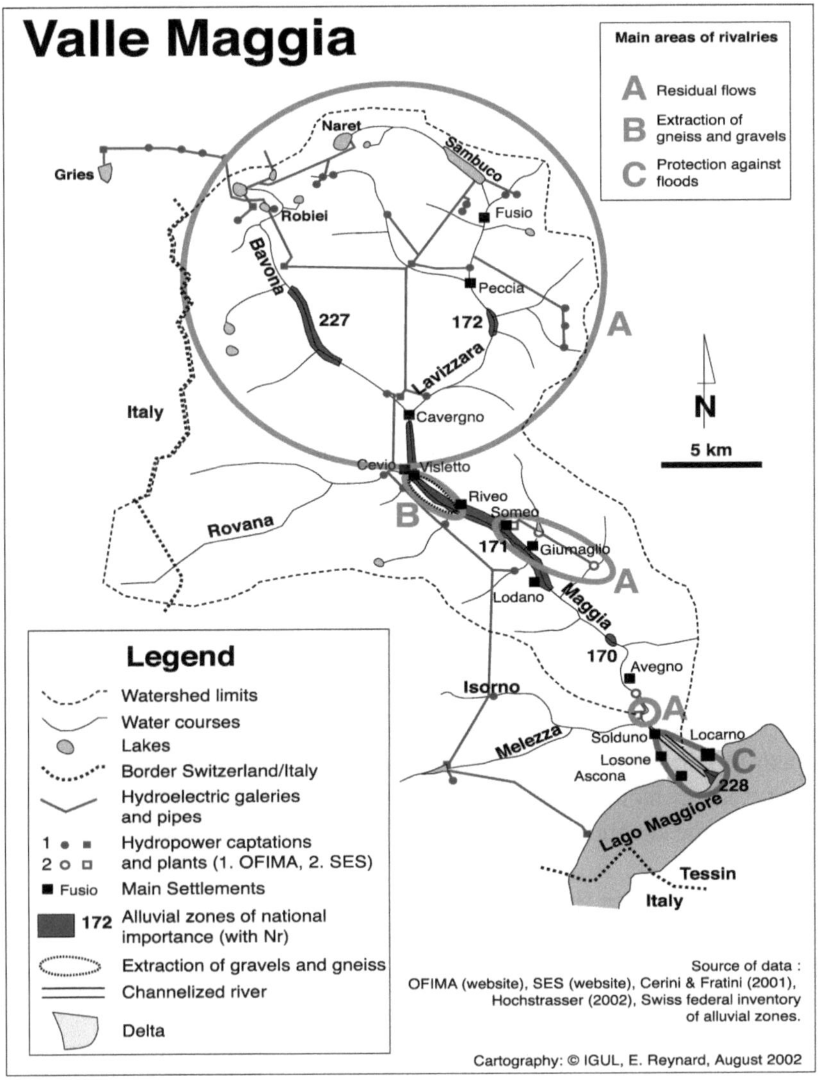

The main water uses in the Valmaggia consist of hydroelectric power production, transport of sediments, support for recreational activities, and natural habitat.

In terms of *hydroelectric power production*, two hydroelectric companies operate using the waters of the Maggia. The private company SES (Società elettrica sopracenerina SA) carries out its sampling in the lower part of the valley and has two relatively small plants. The second much larger company OFIMA (Officine idroelettriche della Maggia SA), a joint venture with 20% owned by the canton of Ticino, runs six power plants. It releases the waters tapped in the upper part of the Valmaggia directly into Lake Maggiore which causes a significant drop in residual water flows over the whole course of the river. Since the advent of hydroelectric power production here, the Maggia has lost about three quarters of its original water volume.

With respect to *geomorphological processes and natural risks*, the flow of the Maggia can increase dramatically. The flood of August 7, 1978, following exceptionally violent precipitation, affected the whole region, causing catastrophic damage and costing many lives.

The most important *recreational activity* in the Valmaggia is fishing. Other recreational activities include canyoning and canoeing, diving from the rocky cliffs and swimming, and also hiking. With respect to *natural habitat for plants and animals*, several zones have been listed in both cantonal and federal inventories for their ecological value. The Valmaggia is the site of two *quarrying* operations: for *gravel and granite*. The annual average amount of gravel taken from the riverbed is 75,000 m^3 over the last ten years. The granite quarries are mainly located in a limited area between Riveo and Visletto on the two rocky slopes on either side of the river, but the stone-processing operations take place on the banks where the largest plants and logistical structures are located. The annual yield from the quarries on the Maggia is in the region of 70,000 m^3 of natural stone. The waste (35,000 m^3) is partly used in the re-processing facilities in the quarrying area or it is dumped, in particular around the edges of the river bed. Other uses for the water resource include the consumption of drinking water, as well as provision of the strategic reserve for fire fighting and irrigation.

8.3.2 Three main fields of rivalry

The main rivalries in the Valmaggia related to the water resource during recent decades concern hydroelectric production, gravel and granite quarrying activities, and protection against floods.

Hydroelectric production

In 1965, the first signs of damage to the hydric system and to the landscape due to the production of hydroelectric energy became noticeable in the Valmaggia. The water was extremely scarce, especially in summer, and the groundwater level fell, creating problems for drinking water suppliers. The main reaction came from fishermen. After the completion of several studies aimed at defining minimum residual flow rates acceptable to the companies to avoid claims for compensation, the canton and OFIMA entered into a provisional agreement between 1969 and 1973 and the first minimum residual flow rates became effective, without, however, any specific basis in law. Nonetheless, the public and political debate continued. Cantonal parliament members denounced the excessive use of the water resources to the detriment of the local population citing the 'energy barons' as the only beneficiaries. In 1975, the Fisherman's Federation launched a Popular Initiative to introduce a new article into the *Cantonal Law on the Use of Water* which would force the canton to guarantee sufficient minimum residual rates when it grants a concession, taking into account drinking water needs first and foremost, but also fishing, the protection of the water and the natural environment. The Initiative passed with ease in December 1976.

From then on it was a question of accurately defining the residual water flow rates to be imposed on the ground. With OFIMA's agreement, the Council of State (executive) proposed increasing the minimum residual water flow rates by 1%, without compensating the hydroelectric companies. However, in October 1982, the parliament instead passed a statutory order setting the increased residual water flow rates at approximately 2% for current and future concessions. All concessions were then re-specified and the residual water flow rate modified for each catchment. September 1983, saw OFIMA submit a legal action in the Federal Courts demanding compensation from the canton. There ensued a legal dispute which has been settled by a judicial agreement finalised before the Federal Court in June 1996. In this agreement, the canton acknowledges that the reductions imposed in 1982 were part of the watercourse rehabilitation measures, which anticipated the measures provided in the new *Federal Law on the Protection of Water* of 1991. The corresponding loss of production for OFIMA was 2.4%, for which no compensation was provided. Nonetheless, compensation will be paid for any additional rehabilitation measures.

The *Federal Law on the Protection of Water* imposed a new minimum residual water flow rate in the early 1990s. This new minimum is very low (minimum of 50 litres/second), but is compulsory for all watercourses and must be imposed when concessions are renewed. In regions which are part of the national inventories that list areas of high ecological value, yet further rehabilitation measures can be demanded. Such negotiations are taking place between the canton of Ticino and the hydroelectric company, SES, currently to redefine the parameters of the concession for the Avegno tapping stream. The initial attempt to reach agreement encountered opposition from the WWF and the Fisherman's Federation, which are demanding accurate calculations of the minimum residual water flow rates.

Gravel and granite quarrying

As far back as the 1980s, the Rivo-Visletto sector drew the attention of the canton of Ticino due to the problem of the mineral waste deposits from the granite quarries. This debris had been deposited directly at the edge of the Maggia, without any authorisation whatsoever, and in contravention of the existing legislation in force for the protection of the riverbanks. Nevertheless, for quite some time, at cantonal level, policy has provided a regulatory framework for the conflict between gravel and granite quarrying and other uses. Finally, during the 1990s, this 'greening' of policy intensified with the *Federal Law on the Protection of Water*, which underlined the importance of geomorphological issues over mining activities, the *Inventory of Alluvial Sites of National Importance*, which designates the Riveo-Visletto as a site of ecological value, and the federal and cantonal legislation on fishing, which protects the natural habitat of the fish. In addition, the canton regulates quarrying in the framework of the *Cantonal Law on Building*, which requires a detailed planning authorisation and a plan for restoring the area for all mining or quarrying activities. In short, the complete process of developing a more environment-friendly policy should have resulted well before the 1990s, but in reality, in the Riveo-Visletto sector, authorisations were granted for decades in a more or less informal fashion and without any related requirements.

In 1997 the Someo town council, worried about the situation, requested a meeting with the director of the cantonal planning department. Aware of the gap mentioned, he commissioned a firm of engineers to conduct an in-depth survey of this sector to assess the environmental impact of past and present operations. In fact, with the reports completed in 1998 it became evident that the problem of granite waste deposits is not the only problem in this sector. Other issues to contend with are gravel quarrying, which proved to be excessive, as well as the protection of listed alluvial zones. These conflicting interests are concentrated in a small area and are intimately interlinked. The reports illustrate that the various activities existing in parallel are marked by an absence of co-ordination, both on the part of the local authorities and the companies involved. The two reports are implacable in their conclusions: the ecological situation in the area is alarming. The vast majority of current authorisations, generally covering short periods of one year, had been suspended by the canton in the interests of the general public at the end of the 1990s. It now remains to find a solution for terminating this activity that will not spell economic disaster for the region.

To date, there are several solutions under negotiation. This follows a process of consultation extended to all actors involved in a climate of reciprocal co-operation. Re-processing and grinding of the quarry waste to make it re-usable in the place of the gravel quarried from the river is one of these ideas. There is now an agreement between the gravel and granite quarry operators and the procedure seems to be working well. Finally, the gravel-quarrying sector could be used in the works related to hydraulic safety.

Protection against floods

During the 1980s, plans were drawn up for a complete reclamation project for the Maggia, but due to the sparsity of population in Valmaggia, the project was put at the bottom of the queue. When the project was reviewed, the problem of minimum residual water flow rates and an awareness of the environmental importance of the river took precedence and the plans were finally abandoned.

There is a long history of flooding in the lower part of the valley, affecting the area of Locarno first of all, but also reaching Ascona. Despite extensive precautions such as channelling and widening the deltaic river stretch of the Maggia during the previous decades, the catastrophic flood of 1978 devastated the Locarno deltaic stretch. To prevent catastrophe on this scale, the area needed to re-design and re-build the protective infrastructures. Whereas it is true to say that nobody questioned the need for the subsequently planned protective structures, the new project has not met with equal enthusiasm among all the local people. In fact, if policy design has long promoted structures to channel the rivers, especially at cantonal level, it nonetheless gradually integrated competing uses and provides a basis for safeguarding the natural habitat and maintaining the natural river dynamic. 1978 saw the recognition of the need to maintain the riverbanks as a natural habitat, in both federal and cantonal regulations.[7] At that point, environmental organisations were critical of the linearity imposed on the river and the degrading of its banks caused by the very geometric stepped bank built along it, especially where the natural habitat could still be conserved. A compromise was finally reached with the building of more natural mosaic-style banks. This can be interpreted as an arrangement between the informal 'use rights' to the flora and fauna, defended by the WWF, and the need for hydraulic safety, for which the canton is responsible and which unites all actors. Work began on this stretch in 1979 and was completed in 1982, without any further objections.

The second section, from the Solduno bridge to the lake, was more problematic. The municipality of Locarno was especially concerned about the impact of the structures on the landscape. It also disputed the height of the safety bank. Furthermore, the WWF opposed felling the large trees and

[7] Federal law on the Protection of Nature, the Landscape and Cultural Heritage, Regulation in application of the Statutory Order of January 16, 1940 on the Protection of the Natural Heritage and the Landscape.

even insisted on planting more on the strips of public land bordering the river. It was particularly insistent on protecting the nature reserve it manages at the mouth of the Maggia and is of the opinion that the safety levels demanded by the project are exaggerated. In 1985, a compromise was finally agreed, after seven years of discussions.

However, two years later, the owners of the Delta camp site and of a hotel complex, just behind the natural reserve area managed by the WWF, submitted objections. They stated that their land was insufficiently protected against water. A breach which had appeared in the unsinkable bank on the left bank was discovered and was repaired in 1989. A further demand was that the insubmersible banks be extended on the left bank. The WWF is opposed to this addition to the banks and the municipality of Locarno is divided between these two conflicting demands. In 1995, the suggestion of the construction of banks behind the nature reserve met with approval from both sides of the argument. Work commenced in March 1997 and was completed in early 2000.

8.3.3 Regime change in the Valmaggia

In all three sub-cases we can observe the resolution of rivalries between heterogeneous uses through a gradual introduction of interests of other uses into the regime regulating one specific use between the 1970s and today. With respect to hydroelectric power production, a change in policy design introduced the quantitative protection of the water resources at cantonal level in advance of other legislation as a result of the Fisherman's Federation's Popular Initiative. By this, hydroelectric companies' use rights were restricted in favour of the other functions of the water, i.e. landscape, natural habitat and fishing. In the case of gravel and granite extraction, the policy design increasingly incorporates uses that compete with quarrying. In some respects, the various projects constitute a transfer of the use rights to new areas of use, even if the same companies are involved. They no longer simply produce a construction material, but also recycle the granite waste, thus ensuring the hydraulic safety of the riverbanks. By the restoration work, they are also contributing to renewing the natural habitat. In the field of flood protection, protection against water remains very much to the forefront, but it must accommodate other uses, such as natural habitat or landscape functions. The plans for the development of the riverbanks gradually became more environment-focused as environmentalists fought their corner.

Thus, we can observe an enlargement of the extent of the regional institutional regime and an increase in its coherence due to several co-ordination and integration measures at the substantial level of the policy design (e.g. higher residual flows, prescription of natural river banks in flood protection). On the institutional level of the regime, such a tendency also appears (e.g. co-operation of the various actors involved with respect to extraction of materials) but, however, remains restricted to rather informal strategies (no change in formal administrative structures). All in all, the institutional water regime in Valmaggia has generally developed in the direction of more integration.

8.3.4 Conclusions for the Valmaggia water basin

As to the contribution of these changes in the institutional regime to the *sustainability* of the resource uses in the Valmaggia, we can observe a general improvement. With respect to the ecological dimension, we can fairly safely assert that the introduction of quantitative protective measures for water has had a real impact on the ground. An improvement of the condition of the Maggia was perceptible immediately after informal minimum residual water flow rates were set in place, especially in terms of the ground water level. Nevertheless, their precise determination remains a major challenge and the Maggia is still short of water today.

With respect to gravel and granite quarrying, it is very difficult to draw any conclusions regarding its contribution to the sustainable use of the resource, since actual change at the level of the water basin is a very recent phenomenon. The environmental impact study mentions several criteria. To date, these various conditions have not been fulfilled.

In the third sub-case, the technical compromises achieved as a result of pressure from the WWF have clearly improved the situation from the point of view of ecologically sustainable solutions. In terms of the impact on the socio-economic dimension of sustainability, we may at least state a constant situation, respectively a slight overall improvement for several reasons. First, in cases where compensation is paid out according to federal legislation, this will compensate for the loss of production caused by the limits placed on the hydroelectric companies' use rights. Secondly, in the quarrying sub-case, in fact, transferring use rights enabled the region to avoid an economic disaster, creating a new niche for local businesses. And thirdly, economic and social sustainability was guaranteed for the towns of Locarno and Ascona, seriously threatened by the Maggia's floods. A dyke project that would excessively compromise protection against water in favour of competing uses could not be described as sustainable.

There are similarities across the three sub-cases in terms of *conditions* favourable to fostering change. These conditions include: visibility and evidence of conflicts of use and their negative consequences; existence of relevant scientific information; effective dissemination of information to those involved (especially through the local press, which mobilised public opinion); involvement of organised bodies with legitimate use rights at water basin level; spirit of co-operation between the actors and the incorporation of non-state and other users involved; existence and use of several resources in the Swiss political system; parallel developments in the regimes for other resources (e.g. land or natural heritage); support at federal level for the implementation of change. Without question, the simultaneity of these diverse variables coupled with a change in attitude in favour of protecting the environment enabled changes to be made in the institutional regime governing water towards wider integration at water basin level.

8.4 Fighting against water pollution: the Seetal valley[8]

8.4.1 Water uses in a lake valley in the central plateau

Lake Baldegg and Lake Hallwil are located in the Seetal valley in Switzerland's Central Plateau. Lake Baldegg and its water basin lie entirely in the canton of Lucerne, whereas the southern part of Lake Hallwil lies in the canton of Lucerne and the northern part in the canton of Aargovia (see Figure 8.2). The Seetal valley has experienced a remarkable growth of population in recent decades, mainly at the Lucerne end and in the lower valley municipalities. To date, the Seetal water basin has approximately 24,500 inhabitants. Apart from the regional centre Hochdorf, where industry and trade activities dominate, the Lucerne part of the water basin is mainly characterised by intensive agriculture.

The total surface of the water basin of the lakes is approximately 143 km^2.[9] Four-fifths are situated in the canton of Lucerne (23 municipalities) and one-fifth lies in the canton of Aargovia (10 municipalities). The location of the lakes with the surrounding mountains results in very weak natural circulation of the lake water and hence also a low self-purification capacity. Lake Baldegg is supplied by the river Ron and it drains into the river Aabach. After a river stretch of 3 kilometres the Aabach joins Lake Hallwil and flows out of it at the north end towards its conjunction with the river Aare. Both lakes and their tributaries are supplied by a large number of medium-sized and small creeks.

Under the laws of the cantons of Aargovia and Lucerne and the Federal Swiss state, Lake Hallwil is a public water body. In contrast to this, Lake

[8] This section relies on the Euwareness-report 'Second case study for Switzerland: Lake Baldeggg and Lake Hallwil' by C. Mauch, delivered to the EU in Spring 2002 (cf. also Mauch, 2002a).

[9] With an average of 170 inhabitants per km^2.

Baldegg has been the property of Pro Natura (previously known as the Swiss Association for Nature Conservation, SBN) since the early 1940s and is, therefore, a private surface water body, which in Switzerland is generally the exception.

Figure 8.2: *Overview of the Lake Hallwil and the Lake Baldegg water basin with its main elements relating to the resource water, its various uses, and its institutional regime.*

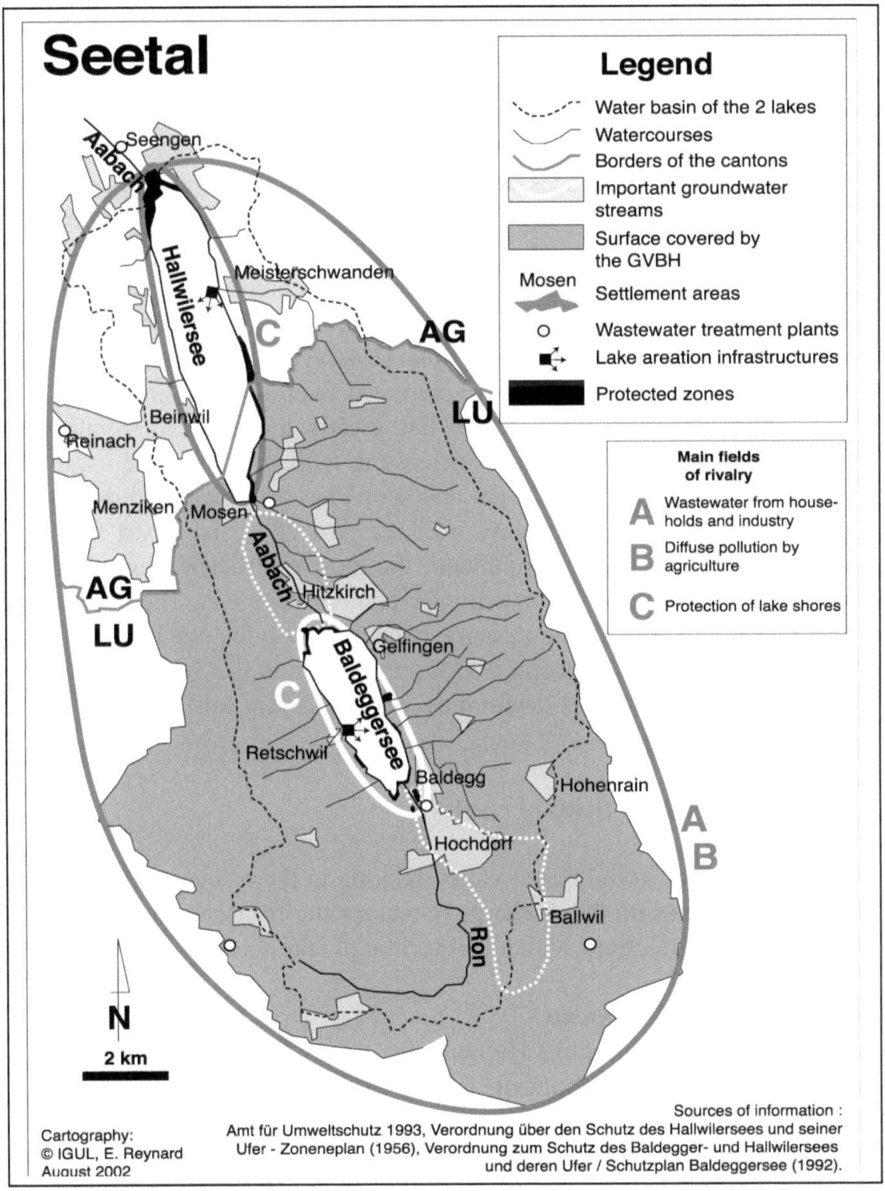

With respect to the goods and services provided by the resource water in the Seetal valley during the past three decades, water as a *living environment* gained significantly in importance, both with respect to surface water bodies, the quality of which improved, and to natural shores and wetlands. Most of these ecologically valuable areas are located at the northern and southern ends of the lakes. In the Seetal valley, in terms of quantity the *drinking water* function decreased during our investigation period. This is the result of both a decrease in consumption and the discovery of large stocks of groundwater in the area between the two lakes. In terms of quality, existing problems with nitrogen in groundwater decreased and levels are now constant (albeit rather high). The importance of *water for economic production* in industry and agriculture has rather decreased due to technical improvements. Despite the difference in the circumstances that prevail today, the function of *transport and absorption* of substances did not undergo a significant change during our investigation period and remains very important. Following the discharge of large volumes of pollutant loads into the lakes in the 1970s in the form of nutrients in wastewater, the situation completely changed in the 1990s. Over a period of around 15 years, pollutant loads from this source were reduced by a factor of almost four. However, within the same period, pollution stemming from agriculture increased fivefold due to intensification in agricultural practices. *Navigation* has greatly increased in significance since the 1970s, mainly due to an increase in tourism and leisure uses. The importance of *fishery* (professional and sport) has also increased due to the improvement of the water quality over the past 25 years. Apart from the good and service 'living environment', the most significant increase in the use of the lakes since the 1970s is for *recreational purposes*. The demand for it and related pressure on the resource is manifest, on the one hand, through the increasing number of people seeking recreation in the Seetal area and in the growing variety of uses (swimming, diving, sailing, surfing etc.), on the other.

8.4.2 Three main fields of rivalry

The main rivalries in the Seetal valley relating to the resource water during recent decades concern the discharge of wastewater from settlements, diffuse pollution from agriculture, and the protection of lake shores.

Wastewater from settlements

Both Lake Baldegg and Lake Hallwil have a long history of pollution. Due to strong economic development, mainly in the municipality of Hochdorf, this trend continued for the first half of the 20th century. In 1940, the whitefish population completely collapsed and this prompted the owner of

Lake Baldegg, a professional fisherman, to decide to sell the lake to a nature protection organisation. In order to reduce the input of polluting substances into Lake Hallwil from household and industrial wastewater, in the early 1960s the canton of Aargovia constructed a sewage system around the Aargovian part of the lake and a wastewater treatment plant in Seengen, which, at the time, represented a pioneering development in Swiss water policy. Due to the insufficient success of these measures, in 1976 the Aargovian municipalities launched a petition to the cantonal government requesting that efforts to eliminate the pollution of the lakes also be undertaken in the canton of Lucerne. This resulted in the decision that the two cantons would work together in the future. They commissioned the Federal Swiss Institute for Water Research (EAWAG) to carry out a study and propose measures for the rehabilitation of the lakes. The study stated that, in addition to the existing practice of promoting the construction of sewage systems and wastewater treatment plants by gradually increasing federal subsidies, efforts to reduce phosphorus loads through lake-external measures and the implementation of additional recovery measures in the lakes themselves were essential. The technical solution proposed involved the direct introduction of oxygen into deep water in summer (lake aeration) and the forced circulation of the water in the winter. The resulting system, which was known as 'Tanytarsus', was tested first on Lake Baldegg and implemented also in Lake Hallwil after the mid-1980s. Following the construction of five more wastewater treatment plants, the sewage system in the water basin was completed in the early 1980s.

In the canton of Aargovia, the lake rehabilitation measures were from the outset implemented and financed by the canton. The situation in the canton of Lucerne is very different since here, according to the cantonal *Introductory Law to the Federal Law on Water Protection* of 1974, the implementation of water protection policy is the responsibility of the municipalities. The canton initiated the foundation of the Lake Baldegg and Lake Hallwil Association of Municipalities (GVBH) in 1984, of which all of the Lucerne municipalities which lie within the water basin are members. The association's aim is to achieve the rehabilitation of Lake Baldegg and Lake Hallwil. It developed a rehabilitation concept comprising four lake-external measures (information for the population, implementation of measures to eliminate the causes of pollution in built-up areas, peripheral regions and in agriculture) and one lake-internal measure ('Tanytarsus' system) as main points for action. The GVBH's initial activities were financed by the municipalities (approx. 55%), the canton (approx. 42%) and also by contributions from Pro Natura (2-3%).

The new *Federal Law on Water Protection* (GSchG) in 1991 brought several changes to the regional regime. First, subsidies for local wastewater treatment plants, which had been extended since the mid-1970s at federal level, were eliminated in 1998 under the new objectives of the polluter-pays principle. Secondly, Art. 28 of the GSchG of 1991 introduced an obligation for cantons to implement additional measures in the water bodies themselves if it proves impossible to comply with the prescribed quality standards by means of other measures. This regulation actually duplicated what had already been implemented in Seetal at national level. And thirdly, according to art. 62a, which was introduced into the GSchG in 1998, subsidies for lake-external water protection measures in agriculture could be received from agricultural policy resources; thus, the agricultural authorities will now also assume a share of these costs.

Diffuse pollution by agriculture

In the 1970s and 1980s, agriculture was a target group of water policy, mostly in the context of technical measures, in particular the cleanup of slurry pits. In the years following the enactment of the *Federal Law on the Protection of Waters* of 1971, the federal administration passed several regulations relating to agricultural practice. However, most of these regulations were formulated on a voluntary basis and the extent of their implementation was in the hands of the different cantons. After the collection of wastewater and its treatment in the wastewater treatment plants had been more or less fully implemented in the Seetal water basin, due to its exploitation practices and the erosion of surplus nutrients, agriculture remained the dominant source of pollution of the water bodies.

In the case of Lake Baldegg, by the mid-1970s, phosphorus loads originating from agriculture had reached approximately 25% (2.2 t/a); this increased to 82% (11.9 t/a) in the 1990s. In the case of Lake Hallwil, the ratio of phosphorus originating from agriculture has undergone a similar increase. In the context of this evolution of pollutant loads, the implementation of lake rehabilitation measures in the 1980s was accompanied by a concept for so-called lake-external measures, of which the measures concerning agriculture were the most important. From 1986, based on its 'Concept for Protective Measures in the Shore Belt of Lakes Baldegg and Hallwil' and working in collaboration with the cantonal agricultural authorities, the GVBH developed an environmental consultancy for agriculture. In the Lucerne area of the Lake Hallwil water basin, according to cantonal legislation, the measures were the same as those implemented for Lake Baldegg and, here too, they were implemented and partly financed by the GVBH. When it comes to the Aargovian area of Lake Hallwil, however, the situation is rather different. Here, the responsibility for all measures lay entirely with the canton.

In 1992, changes in the *Federal Law on Agriculture* (Article 31b) saw the advent of a new provision for the payment of subsidies to farmers on the basis of ecological criteria. This system has been implemented in the Seetal valley since 1993 and has resulted in the adoption of 'integrated production' methods by approximately 90% of the farmers in the water basin of Lake Baldegg. Five years later, the adoption of Article 76[10] of the *Federal Law on Agriculture* of 1998 offered the federal authorities the option of paying subsidies from agricultural policy resources to promote ecological measures in agriculture in regions where the targets for water quality cannot be achieved through other measures. Since this is the case for both Lake Baldegg and Lake Hallwil, a corresponding so-called "Phosphorus Project" has been under way for Lake Baldegg since 2000 and a similar project for Lake Hallwil started in 2001. The projects consist of the setting up of agreements between the canton (in the case of Aargovia) or the GVBH (in the case of Lucerne) and the individual farmers with respect to restrictions on uses (e.g. fertiliser use) in specific areas. Thus, the farmer is paid a certain sum per area and year for reducing the intensity of land use and hence decreasing the dispersion of nutrients from the soil into the water. In the case of Lake Baldegg and the Lucerne part of Lake Hallwil, these measures are implemented by the GVBH, for the Aargovian part of Lake Hallwil, the agricultural section of the cantonal authority acts on instructions of the environment division. The lake-internal and lake-external measures led to a gradual improvement of the state of the resource but not, however, to its complete recovery.

Protection of lake shores

The shores of both Lake Baldegg and Lake Hallwil were placed under protection already at an early stage and, as a result, they were largely preserved in their natural state. At the time, the main aim was to protect them against construction. In the canton of Aargovia, an initial *Decree on the Protection of Lake Hallwil and its Shores* passed in 1935. Its success can be measured by the fact that, today, 75% of the Lake Hallwil is surrounded by natural shore. The decree was replaced by a new one in 1956. This defined protected areas with varying protective provisions on and around the lake. Strips of shore of between 10 and 50 metres in width were defined as restricted zones in the entire shore area of the Aargovian area of Lake Hallwil. This was further encircled by a protection zone belt ranging from approximately 200 to 700 metres in width. These provisions were, however, not binding on the private owners of the land. In the Lucerne area of the Seetal, two decrees -- one on the protection of Lake Baldegg and its shores and one on the protection of Lake Hallwil and its shores -- were enacted in

[10] Introduced to the GSchG of 1991 as Art. 62a.

1961 and 1962. Thanks to this initial legislation and the private ownership of the lake by a nature conservation organisation, the shores of Lake Baldegg also remained largely free of buildings. A 1974 report by the planning delegate of the Federal Council described it as the lake with the greatest proportion of open shoreline (i.e. not built on) in all of Switzerland. In the mid-70s, both Lake Hallwil and Lake Baldegg were included in the federal inventory of landscapes of national importance.

In 1986, a new Aargovian Decree on the Protection of Lake Hallwil replaced the 1956 decree, where landscape protection issues were intensified. The fourth part of the Law on Regional Planning, Environmental Protection and Construction of 1993 represented the introduction of proper nature conservation legislation and a new order for the regulations governing water bodies. These sections were produced under pressure of a popular initiative in the area of nature conservation. In 1996, it built the basis for the new Direction Plan which defined a Lake Hallwil Special Area encompassing the Aargovian part of the Lake Hallwil water basin where the cantonal government is implementing its external measures.

In the canton of Lucerne, the nature conservation organisation then known as SBN compiled an inventory of wetlands which are worthy of protection in the mid-1970s. These were included in the federal inventory compiled in accordance with the *Law on the Nature Conservation* of 1966 in 1977, albeit in a slightly reduced form. The zones form a belt of 20 m to 1 km in width around Lake Baldegg.

In order to reduce the discharge of nutrients and contaminants from the lake shore areas, in their 'Concept for Protective Measures in the Shore Belt of Lakes Baldegg and Hallwil' of the mid-1980s, the association of municipalities defined four zones (A to D) with varying risks of nutrient erosion and use instructions of varying stringency (and the corresponding compensation). Zone A is a nature conservation zone in accordance with the cantonal *Decree on the Protection of Lakes Baldegg and Hallwil and its Shores* of 1992.[11] For the ban on fertiliser use compensation is paid in accordance with the cantonal nature conservation legislation, however, based on an agreement between the individual farmers and the canton. Zone B (restricted fertiliser use) is based on agreements between the individual farmers, the local municipality and the GVBH. CHF 10 per m^2 is paid in compensation for the resulting reductions in yields, financed from the budget of the GVBH. No subsidies are paid for Zones C and D as reduced yields are not expected.

The Lucerne cantonal direction plan of 1986 contains the provisions for Lake Baldegg that the entire immediate area around the lake belongs to the

[11] Under the terms of this decree, the area was divided into water, nature conservation, landscape protection and leisure zones with varying use regulations and restrictions.

landscape protection area. The regional planning for Seetal is allocated the task of developing a special plan and concept for the lake and its surroundings. The path around Lake Baldegg is one of the topics dealt with in the cantonal direction plan for walking/hiking paths.

Thus, in the area of shore protection, the policy design in the Seetal of the past three decades did not give rise to any extension of the protected areas. Instead, the trend was to intensify the protection of existing areas. However, a shift from protecting shores from being overbuilt can be observed to protection from other uses, mainly due to the growing significance of recreation (landscape, leisure, sport) at Lake Hallwil and increasingly also at Lake Baldegg which sharply intensified the rivalry with the water as living environment over the course of the study period.

8.4.3 Regime change in the Seetal valley

The most important conflict between uses in the Lucerne and Aargovian areas of the Seetal appears to centre on the antagonism between water as a living environment, on the one hand, and its function in the transport and absorption of pollutants on the other. Thus, this conflict between different uses is a *qualitative* one.

In all three sub-cases, the rivalry between different uses has increased (for the most part) significantly over the past three decades. Regulations have been adopted which primarily targeted a specific rivalry situation. With respect to the co-ordination of levels and scales, our investigation period saw the advent of certain efforts to take the water basin into account in its entirety. This took the form, first, of the foundation of the GVBH which covers the Lucerne part of the Seetal water basin as a whole and, secondly, the development of the Phosphorus Projects for Lake Baldegg and Lake Hallwil early in the 21st century. In this context, federal legislation obligates the cantons to define an area covering the entire basin of a water body as a condition for the receipt of subsidies for further measures aimed at fulfilling the quality standards defined for water in the federal legislation. However, compared with certain developments at EU level, consideration of water basins in their entirety still appears to be very weak in Switzerland to the present day.

The extent of the regime has been enlarged in the Seetal valley mainly due to the reinforcement of the use of the resource as a living environment, the inclusion of farmers' activities as a target group in association with restrictions on use rights and stricter regulations regarding recreational uses of lake shores. The external coherence between the regulative system and the policy design has been increased, first, due to the identification of farmers holding (indirect) use rights to water (transport and absorption) as target groups, secondly, the obligation to implement rehabilitation measures on the lakes[12] and, thirdly, the reinforcement of restrictions on use rights to the owners or the users of lake shores in the interest of nature and water protection objectives. With respect to the internal coherence in the regulative system, there has been only a slight increase in the case of Lake Baldegg due to the admission of Pro Natura (owner of the lake) to the GVBH, albeit in an advisory capacity. The land buying policy of Pro Natura has also brought a slight increase due to a further overlapping of the lake owner and the holders of use rights to water (transport and absorption). The internal coherence in the policy design also appears to have experienced a certain increase, mainly due to further co-ordination efforts between nature and water protection policies and other uses (e.g. recreation, agriculture), on the one hand, and between the two cantons, on the other. Formal administrative structures have, however, not been fundamentally changed into an integrative direction and the three traditional branches of water policy still exist more or less separately in both cantons. All in all, the extent and coherence of the regime, have both undergone an increase in various dimensions. Thus, the regime at the level of the Lake Hallwil and Lake Baldegg water basin moved towards greater integration in the course of the study period.

In the Seetal valley, public policies appear to rely very heavily on public (in contrast to private) organisations and resources for policy implementation. In all three sub-cases, the policies are largely implemented by the cantons or municipalities, even if they are organised in the form of an association of municipalities as in the case of the GVBH. The regulatory capacity of the property rights holders appears to be rather subordinate in the regime development, since Pro Natura, as the owner of Lake Baldegg, has very little influence on the management of the uses which pollute the lake or harm its natural location in any other way. On the contrary, it exerts some influence through property rights to land as opposed to water.[13]

In the area of qualitative water protection, throughout and beyond the study period, the debates and conflicts always centred on money. In the first phase of the promotion of the construction of domestic sewage systems and wastewater treatment, due to the inadequacies in municipal implementation, the federal subsidies for sewage projects and wastewater treatment plants were gradually increased and extended to other activities (cf. Mauch et al., 2001; Reynard et al., 2001). This finally promoted widespread implementation. The situation with respect to measures in the area of

[12] In the case of Aargovia, the owner of the lake Hallwil and, in the case of Lucerne, the holders of use rights to polluted water, i.e. the municipalities.

[13] Policy of buying ecologically valuable plots around the lake, becoming involved in organisations for the improvement of land through the ownership of land within the relevant perimeter.

agriculture is similar. In this case, the first step involved the cleanup of slurry pits. In the following phase, voluntary incentives, initially simple advice, were quickly followed by payments for special ecological services.[14] These (financial) incentives proved to be too low in the Seetal to bring about effective improvements within a sufficiently short period. They were subsequently increased, at national level with Article 62a *GSchG* and in the Seetal area with the launch of the phosphorus projects based on this legislation and also with a project initiated by Pro Natura which, in accordance with the federal legislation, increased the subsidies by half within the framework of agreements with farmers.[15] As all of these measures are based on voluntary participation, it can be assumed that the use restrictions arising from altered agricultural practices are compensated with the payments. From a property rights point of view, this phenomenon can be interpreted as the state purchase of use rights to water from the respective holders of the rights in order to protect water quality.

8.4.4 Conclusions for the Seetal water basin

The evaluation of the regime change and its effects in each case is intended to be based on three economic, ecological and social indicators for sustainable use. Even if it was not always possible to find reliable 'hard' data (especially regarding the economic and the social dimension of sustainability), we were, however, able to identify some trends regarding these two dimensions in a rather qualitative way which also allow a certain judgement of the overall trend in the Seetal. The evolution of the indicators[16] reveals a tendency towards greater sustainability in the uses and in the state of the resource in the Seetal valley from the 1970s to the end of the last century.[17] The best sustainability performance was observed in sub-case 1 (wastewater from settlements), and the least improvement appeared in sub-case 3 (protection of lake shores) which, however, started from a good state already at the beginning of the investigation period. We can, however, observe some variation between the different dimensions of sustainability, since its *ecological* and the *social* aspects reveal the clearest improvement, whereas the *economic* aspects show more indicators with a decrease and,

[14] According to Art. 31b of the *Law on Agriculture.*

[15] The project was co-financed by Pro Natura, the Swiss Agency for the Environment, Forests and Landscape (BUWAL) and the Swiss Landscape Foundation.

[16] For further details regarding to the indicators cf. Mauch 2002a and Mauch 2002b.

[17] Out of nine indicators per sub-case (three for each dimension of sustainability, i.e. ecological, social and economical), the three sub-cases show an evolution towards more sustainability in 17 dimensions, a deterioration in 4 dimensions, in 3 dimensions the evolution was neutral and in 3 dimensions no reliable information was available (Mauch 2002a, Mauch 2002b).

here, improvement[18] and deterioration[19] more or less equalize. The extent to which this development can be explained entirely by the regime evolution observed for the three sub-cases remains open.

The main *trigger for regime change* in the Seetal appears to be the perceived problem pressure, i.e. the state of the resource. In all three sub-cases we were able to identify an impact from specific characteristics of the Swiss political system, mainly the possibilities of intervention through the instruments of direct democracy (petition, popular initiative). In fact, the owner of Lake Baldegg, the nature conservation organisation Pro Natura, could only influence the evolution of the regime and the measures to protect its property, which was threatened by pollution, as a non-public pressure group in the policy system and not due to its property status.[20] The adoption of various measures in the different fields of regulation related to the resource led to an increase in the complexity of the regime. In a later phase, attempts were undertaken to establish the integration of the regulations relating to the different uses. These appear to have been mainly promoted by the top-down impact from federal legislation and the change in the national determinants of the water regime. However, the particular case of the Seetal also shows some bottom-up trends in terms of impacts from local approaches to the solution of problems[21] to the subsequent federal regulations. Integration at the level of the water basin was somewhat forced by the canton of Lucerne and only occurred later in the canton of Aargovia on the basis of the new agricultural policy requirements.[22]

8.5 Analysis of the Valmaggia and the Seetal valley cases in their context

The following section first compares the two case studies regarding to central dimensions of the analysis of institutional regimes and, secondly, draws some conclusions on the impact of the change in the national regime determinants on the regional institutional regimes in the Swiss case studies.

18 E.g. water consumption per inhabitant, offer of tourist infrastructure and activities.

19 E.g. Costs of wastewater treatment plants and the rehabilitation of the lakes, costs for protection of restricted areas.

20 Apart from its policy of buying land around the lake, which is a policy concerned with the resource soil rather than water.

21 Rehabilitation measures on the lakes, offer of advisory services to farmers by the institutions responsible for water protection at regional level.

22 Obligatory definition of the water basin area in its entirety and development of a strategy for this whole area according to Art. 62a of the federal GSchG.

8.5.1 Comparison of the Valmaggia and the Seetal cases

The Valmaggia and the Seetal case studies vary in several respects. First, their geographical location in Switzerland is very different. The Seetal represents a typical Central Plateau valley, a region which was directly affected by the population growth, industrial and societal development of the past decades. In contrast, the Valmaggia stands for a steep alpine valley with very low population density and modest economic productivity where mainly tourism has evolved significantly. In parallel to these different geographical, social and economical structures, the uses of the resource water also vary. In the Valmaggia, its use for hydropower production dominates, having emerged at the end of the 19th and being further developed in the second half of the 20th century. Considering the fact that the Maggia is a mountain river which, greatly depending on the precipitation in the area, can evolve into a furious torrent within a very short time, protection from floods also becomes a very important feature in the region of its river mouth, a plain area (rare in this region) suitable for human and industrial settlements. In contrast, these two uses are of relatively little or even no importance in the smooth Seetal valley landscape where intensive agriculture and settlements dominate, activities which become a heavy burden to the two standing waters with low circulation capacity, i.e. Baldegg and Hallwil Lakes. Even if, related to the general socio-economical development with growing demands in terms of number and heterogeneity, in both cases an increase in the extent of water uses can be observed over the past 30 years, rivalries still have evolved differently. However, in both cases they centre around the characteristics of the main uses in the region. For the Valmaggia this is its quantitative aspects and for the Seetal, it is its threat in terms of quality.[23]

At the level of the actors and institutions involved in both cases, the local population makes use of instruments closely related to the specific Swiss political system. In the Valmaggia, fishermen and local people fight against too low residual flows in the rivers by means of a popular initiative, and in the Seetal the local population in Aargovia urges their government to cooperate with the canton of Lucerne by means of a petition. In the Valmaggia and the Seetal as well, such activities emerge in the context of, firstly, the most fierce conflict on water uses in the region and, secondly, two cases of scarcity which become obvious and perceptible to the population (dried up rivers in Ticino, fish kills and overproduction of algas in the Baldegg and Hallwil lakes). In both cases, under this 'bottom-up' pressure, protective measures standing in opposition to economic uses (hydropower

[23] This difference was taken as a selection criterion for the case studies in order to cover various use and rivalry situations which are relevant in Switzerland.

production, gravel and granite quarrying, industrial and agricultural production) were introduced by state authorities, be it the cantons or -- subordinate to their institutional and regulative framework -- the municipalities. We can observe a process of acting on existing use rights on water[24] through policy measures in all six sub-cases. In this process, existing use rights are gradually newly defined and are later again 'stabilised' on another level.

In both case studies we have encountered administrative institutions at the level of the cantons which are divided along the three main branches of water policy we had already identified at national level.[25] The basis of these administrative structures has not changed in either case over the past decades. However, we can observe the arrival of more and more steps towards co-ordination and integration efforts in specific problem situations. In the Valmaggia this was the case with the establishment of an interdisciplinary working group dealing with the quarrying problem case and in the case of the solution of the flood protection measures in the Maggia delta. Such co-operation attempts also arose in the hydropower sub-case due to a judicial agreement finalised before the Federal Court in 1996, where differing interests were balanced. In the Seetal case, such co-ordination and integration efforts seem to have emerged less often, but are now, however, of a certain importance in the field of water protection from diffuse pollution by agriculture and also in lake shore protection.[26]

We may, therefore, state that at the level of policy implementation *in situ* (i.e. substantial aspects of the policies) we can find increasing activities of integration between rival uses, but, however, at the institutional level these changes do not really seem to be backed up, and traditionally separate institutional structures persist. Nevertheless, we would like to stress the possibility that -- in contrast to the national level where the Swiss political system is well known for its slow processes -- at local level, actors might act more rapidly, i.e. adapt (local) institutions and regulations more quickly to new problem situations (e.g. new rivalry situations between different uses of natural resources).

With respect to the impact of the observed regime changes on the sustainability of the water uses, the Valmaggia and the Seetal case as well show a clear improvement in the ecological dimension. In both cases, the improvement in terms of social and economic sustainability, even if also

[24] Water concessions for hydropower production, quarrying concessions, "right to protection of areas from floods", wastewater discharges, storage capacity for liquid manure and over-fertilisation in agriculture causing diffuse water pollution, free use of lake shores.

[25] I.e. protection from floods, hydropower production, water protection.

[26] Activities and negotiations in the association of municipalities GBVH; weighing of interests in the regional planning group regarding to tourist infrastructures, especially the hiking path around Lake Baldegg.

present, show less clear evidence. One interesting finding with respect to the different ownership structures to the resource results from the fact that we were not able to identify any clear difference in the impacts on the sustainability of the resource uses between the case of private ownership of Lake Baldegg and the other, publicly owned, water bodies (Lake Hallwil and Maggia river). This result appears to stand mainly in relation with a specific characteristic of the resource water, i.e. its dynamic character. Private ownership of Lake Baldegg does not prevent its owner, Pro Natura, from having it polluted, be it by wastewater from settlements, or, more acute in the recent decades, from diffuse pollution by agriculture. Due to the dynamic and moving nature of this resource, its (polluted) parts enter Lake Baldegg through the hydrological cycle and, hence, one central characteristic of private ownership, i.e. its exclusivity respectively the exclusion of other actors than the owner from its use, can actually not be maintained.

In terms of conditions fostering regime change, the comparison of the two Swiss cases draws the following picture. In both cases, Valmaggia and the Seetal valley as well, the visibility of rivalry respectively its negative consequences, the existence of relevant scientific information, the involvement of motivated people at the water basin level, the use of instruments of Swiss direct democracy, parallel developments in other resources' regimes towards more environmentally friendliness, and -- at least at a later stage -- the support at federal level, proved to promote regime change in the direction of integration. We can, however, observe at least one rather striking difference in the process of regime development in the two cases. Whereas in the Valmaggia, a spirit of co-operation between the actors and the incorporation of non-state and other users involved in the conflict in appropriate structures appears to be an important issue, we must, in the case of the Seetal valley, state that this was not the case. On the contrary, in the Lucerne part of the water basin, for example Pro Natura, even if it is the owner of Lake Baldegg, seems to have been systematically excluded from the association of municipalities (GVBH) committee for a long time, a situation which now appears to be relaxing only very slowly and in very small steps. In the canton of Aargovia, mainly due to the single responsibility of the canton as the lake owner for the lake recovery measures, other actors did hardly appear on the scene.

8.5.2 Evolution of the national determinants and of the regional institutional regimes: extent and coherence

During the past century, at national level especially, the *extent* of the goods and services regulated has been enlarged. In a first phase (last decades of the 19th and first decades of the 20th century), this enlargement has mainly occurred in the field of exploitation uses (protection from floods, then hydropower production). Furthermore, it was significantly property rights driven (e.g. assignment of use rights to hydropower producers through concessions). At its later stage, the enlargement has mainly occurred through protective issues and was policy driven.

This increase in the extent at first hand led to an increase in the differentiation of the regulations and, hence, a decrease in the *coherence* within the regime determinants. Even if we can observe a growing interrelation between different water policy fields[27], this co-ordination appears to remain more or less on a substantive policy level, and does not come into action to the same extent on the institutional level.[28] The fact that in Switzerland the institutional arrangements are generally implemented by the different cantons (which assigns them a great importance in the policy design) may serve as an explanation here. The Confederations capacity to generate coherence, therefore, remains low within the Swiss federalist political system.[29] The *internal coherence in the policy design*, therefore, has to be judged rather low (low on the level of institutional aspects, even though high on the level of substantive aspects). With respect to the regulative system, in practical terms 'unclear' definitions (*Unschärfen*) occur in the public policies and in the related definition of property, disposal and use rights. In this respect, for example the prescription in the water protection law of 1991 which urges the cantons to rehabilitate water courses, from which large quantities are drawn and where residual flows are not guaranteed, even before the end of the concession period as far as no duty for compensation arises, represents an unclear definition, since the law does not define from what amount of reduction of water quantities this holds true. Thus, the *internal coherence of the regulative system* tends to be low at the national determinants' level, at least at to the beginning of a policy cycle.

However, at regional level, e.g. agreements concerning the definition of residual flows could be found in the Valmaggia at the time, and, hence, the situation with respect to coherence generally appeared to be more favourable in the case studies, mainly in the context of specific regional problem solution strategies. Thus, the situation regarding institutional regimes at regional level appears to differ from the national determinants' level. We might interpret this difference as 'temporary coherences' emerging at regional level, i.e. specific local regime arrangements centred around

[27] I.e. between exploitation and protection policies (e.g. with the instrument of the General Water Discharge Plans (GEP) in the water protection law of 1991).

[28] This is an observation which particularly results from the two case studies, Valmaggia and Lakes Hallwil and Baldegg, and could not at first hand have been expected on the basis of the national screening.

[29] In the European context, this appears to be unlike to the situation e.g. in France and the Netherlands but similar to the Italian situation.

specific local problem situations which reach beyond the structures given from the national level. We, therefore, can observe the flexible emergence of '*unstable integrated regimes*' at regional level which succeed in producing (temporary) integrated institutional regimes focusing exclusively on the solution of concrete problem situations.

8.6 Conclusions: 'problem basin approach' in Switzerland

In the next section, several conclusions are drawn from the analysis of the case studies with respect to interactions between the public policy and the property rights dimensions of the institutional regimes, the emergence of institutional arrangements in the policy design along rival uses, and possible explanations promoting change towards more integration in the regime in Switzerland.

8.6.1 Interaction between the public policy and the property rights system

At the demarcation line between a legitimate 'concretisation' of water uses (e.g. limitation by a public policy), where no compensation payments are owed, and the situation of legally accepted material expropriation with financial compensation, the public policy and the property rights system interact. Up to this line, users of a resource are urged to tolerate the limitation or decrease of their use rights (for example) in the public interest. Beyond this line, they have a right to compensation.

The process of interaction between the public policies and the property rights system might be described as follows: with the arrival of a new public policy, use rights to the resource are newly defined. A debate starts on the limit to a 'real' material expropriation, i.e. where compensation payments are owed. Once the new structure is institutionalised by the new public policy, further limitations may come into force some years later through a court decision restricting use rights even further.[30]

Thus, at the demarcation line between property rights and public policies the property rights system is 'mobilised'. This is due to the fact that every

[30] This limit to 'real' material expropriation, e.g. with respect to the rehabilitation of watercourses according to article 80 of the Swiss Law on Water Protection of 1991, has not yet been generally determined, neither by court decisions or by a legal doctrine. Experts in general demand for a judgment of every single case. However, in any case they consider productivity losses up to 4% to be free from a need for compensation (cf. Eckert, 2002: 152).

intended impact of any public policy related to the water resource needs to influence the behaviour of the actors actually making use of the resource in order to change it. Hence, with the emergence of every new good and service regulated, the limits between the public policies and the property rights system respectively the definition of the property rights system in terms of concrete use rights (rights to make use of the resource in this or another way) has to be newly defined. At a later stage, this *destabilisation of the property rights system* first leads to a new stabilisation on another level and than to a new clear-cut definition of use rights.[31]

In Switzerland we have identified a phase of high extent of the institutional regime (IR) from the beginning of the 90s onward, but we can not actually identify a convincingly high level of coherence. This is, on the one hand (regulative system), due to the fact that unclear definitions will exist up to the moment where e.g. a court decision on the limit up to where the use rights may be restricted without a need for compensation is taken. On the other hand (public policy), it is a consequence of the specifities of the Swiss political system (strong position of the cantons which define most regimes at the regional level). Therefore, the national determinants of the regional regimes tend to give room for highly varying institutional regimes at regional level -- a tendency which we assume to be clearly stronger for Switzerland than for the other European countries due to the specific federalist Swiss political system. Furthermore, with respect to the fact that in Switzerland changes in property respectively use rights on water have generally been implemented through public policies and not through the regulative system during the past decades, we might assume that here, the property rights system appears to be 'stronger' respectively more stable than the public policies due to a traditionally strong position of property rights in Switzerland. Regarding the comparison of the six country cases, we would for the same reason state that the destabilisation of the property rights system through the arrival of public policies appears to be stronger in Switzerland than in the other countries.

8.6.2 Institutional arrangements along rivalrous issues

Both the country screening and the case studies, have revealed that in Switzerland institutional arrangements in the water policy field are strongly organised around and, hence, separated along the three different traditional fields of water policy, i.e. flood protection, utilisation of water (mainly hydropower), and water protection. As a matter of fact, these three branches focus on the main rival uses of water and are furthermore strongly mirrored

[31] This also means that, in a first phase of the arrival of a newly regulated good or service, the internal coherence of the property rights system will be newly defined.

in the institutional structures of these policy fields. They actually form three different 'policy communities'. These institutional structures appear to be very persistent and tend only to change, to open up towards each other under very strong pressure of rivalries respectively conflicts between various uses (first at the local/regional level).

This observation forms an obvious contrast to the water basin approach aspired to in the EU member countries. In Switzerland, such an approach could actually not be identified in any of the case studies, respectively only very recent and tentative approaches in this direction towards the end of the 20th century. The approach found in Switzerland rather appears to be some sort of a (virtual) *'rivalry basin approach'*, i.e. the 'basin' is rather formed around existing rival respectively conflict uses than by a geographical area.[32] Another effect of this kind of 'pragmatic' approach seems to be that inter-policy and inter-resource aspects appear to gain a more decisive importance than in other countries due to the focus onto concrete problems in a certain area (e.g. nature protection and water protection). This means that in Switzerland the regimes are formed around specific goods and services which actually exist in a certain area, i.e. they depend on a concrete problem situation. Thus, generally speaking only goods and services which stand in a rival relationship are covered by the institutional regime.

Within the observed bottom-up processes, where local problems are first solved at the local or regional (or cantonal) level and then 'transferred' to the national level,[33] the subsequent 'recognition' of (former) local problem solution strategies through federal legislation offers a further legitimation to these regional regimes.[34]

We might, therefore, state that Swiss water regimes are *integrated by rivalries* ('rivalry basins') rather than with respect to the resource as an entirety. One impact of this, observed in the case studies, is that inter-resource aspects have a stronger position in the regime than might be the case in a situation where the water basin approach dominates.[35] However, the question still remains on the table whether the 'water basin' or the 'rivalry basin' approach will finally guarantee more sustainability of the uses.

[32] E.g. fishermen vs. hydropower producers in the Valmaggia, users of the water for transport and absorption (i.e. settlements and farmers) vs. water protection in the Seetal valley.

[33] E.g. cattle limitation in the Seetal, minimal residual flows in the Valmaggia.

[34] E.g. definition of minimal residual flows in the Federal Law on Water Protection of 1991.

[35] Cf. for example France where the regime is strongly basin oriented but inter-resource aspects do not seem to be important.

8.6.3 Possible explanations

With respect to possible explanations for these observations, we would like to propose three particularities of Switzerland. First, the observed bottom-up effects result from the specific Swiss *federalist system* where cantons often function as 'laboratories' for national solutions. Local problem solution approaches are 'legitimated' through subsequent national legislation.[36] Secondly, the *heterogeneous geographical structure* of Switzerland (Alps, Central Plateau, Jura) leads to very different solutions due to very different problem situations. And thirdly, the *complexity of the Swiss political system* (direct democracy with its instruments) allows for a better acceptance of specific rivalry issues.[37] Generally speaking, these aspects lead to a rather 'unsystematic' way of promoting water regime related topics in Switzerland, a way which appears to be strongly determined by specific goods and services in certain areas.

REFERENCES

Eckert, Maurus (2002) *Rechtliche Aspekte der Sicherung angemessener Restwassermengen.* Schriftenreihe zum Umweltrecht, Band 18. Schulthess Juristische Medien AG, Zürich.

Leimbacher, Jörg and Thomas Perler (2000) *Juristisches Screening der Ressourcenregime in der Schweiz (1900-2000).* Working paper no. 8/9 de l'IDHEAP, Chavannes-près-Renens.

Mauch, Corine (2002a) *Second case study for Switzerland: Lake Baldeggg and Lake Hallwil.* Zürich/Chavannes-près-Renens. (delivered to the EU, unpublished)

Mauch, Corine (2003) Institutionelles Wasserregime im Einzugsgebiet von Baldegger- und Hallwilersee (Seetal). Fallstudie. Working paper de l'IDHEAP, Chavannes-près-Renens

Mauch, Corine and Emmanuel Reynard (2004) *Water Regime in Switzerland.* In: Kissling-Näf, Ingrid and Stefan Kuks (eds.): The Evolution of National Water Regimes in Europe. Transitions in Water Rights and Water Policies towards Sustainability. Kluwer Academic Publishers, Dordrecht.

Mauch, Corine, Emmanuel Reynard and Adèle Thorens (2000) *Historical Profile of Water Regime in Switzerland (1870-2000).* Chavannes-près-Renens: Working paper de l'IDHEAP 10/2000.

Mauch, Corine, Emmanuel Reynard and Adèle Thorens (2003) *Le changement du régime institutionnel de l'eau en Suisse entre 1975 et 1991.* Chavannes-près-Renens: Working paper de l'IDHEAP.

Reynard, Emmanuel, Corine Mauch and Adèle Thorens (2001) *Développement historique des régimes institutionnels de l'eau en Suisse entre 1870 et 2000.* In: Knoepfel / Kissling / Varone: Institutionelle Regime für natürliche Ressourcen: Boden, Wasser und Wald im

[36] E.g. the minimal residual flows existed in Ticino and later were introduced into Swiss legislation; similarly, the cattle limitations per area existed already in the Aargovian and Lucerne Seetal before entering national legislation.

[37] Cf. the Valmaggia case where visible aspects of the quantitative water problems were able to mobilise the fishermen and launch a political process; or the (visible) pollution of the Seetal lakes which resulted in a popular petition. In both cases, public opinion was crucial.

Vergleich - Régimes institutionnels de ressources naturelles: analyse comparée du sol, de l'eau et de la forêt. Reihe Oekologie & Gesellschaft Nr. 17, Basel / Frankfurt:Helbing und Lichtenhahn.

Reynard, Emmanuel, Mauch Corine et Adèle Thorens (2000) *Screening historique des régimes institutionnels de la ressource en eau en Suisse entre 1870 et 2000.* Chavannes-près-Renens: Working Paper IDHEAP 6/2000.

Thorens, Adèle (2002) *First Case Study for Switzerland: The Valmaggia,* Chavannes-près-Renens: Working Paper IDHEAP 4/2002.

Chapter 9

Integrated Governance and Water Basin Management

Comparative analysis and conclusions

Hans Bressers and Stefan Kuks
University of Twente (Enschede-Netherlands)

9.1 Introduction

The research in this book has some specific characteristics. While we did not look only from the perspective of immissions or emissions for the protection of habitats, but took a resource perspective, a greater variation of uses and users was drawn into the analysis. Nor did we not restrict ourselves to a public policy perspective or a property and use rights perspective, but combined the two, both theoretically and empirically.

This chapter presents the results of the comparative analysis. This analysis is based on the assessments of their cases (including subcases) by the researchers of the main variables of the theory used. These assessments were based on ordinal scales with five values. The 24 (sub)cases and 13 variables per case are of course too many to be handled in a purely qualitative way. Therefore the analysis below mostly uses descriptive and analytical statistics that are apt for ordinal level variables. Some of the main conclusions are illustrated by real life examples from the case studies.

9.2 Regime change

What interests us here is the degree to which the listed aspects of the regime, separately and as a set, moved in the direction of more integration (extent and coherence).

Hans Bressers and Stefan Kuks (eds.), Integrated governance and water basin management. Conditions for regime change towards sustainability, 247-265.

The *extent* is the degree of completeness of the domain of the regime in terms of relevant uses and users.[1] In most of the cases and subcases in the study the extent of the water resources regime changed positively, in many cases even to include more or less all relevant uses and users.

The *internal coherence of public governance* is the degree to which the interdependencies in the water system and its management that occur in reality, are reflected within and between the contents of the elements of public governance.[2] The internal coherence of public governance generally increased too, but less than the extent. Almost nowhere could a 'full coherence' statement be made and in several instances only small improvements occurred.

Remaining difficulties with non-river basin jurisdictions

In France the SAGE process has generated a collective dynamic. Among other things the extent of the regime that was slowly built before, was quickly enlarged. The SAGE process could build on the gradually increased openness to cooperation that emerged over the last 25 years. The SAGE procedure has led to awareness of most (and new) stakeholders that they are not the only one 'main' user. But that doesn't always imply that there is participation from all actors or this participation is dedicated to reinforcement of collective action, but rather considered by some powerful users as a way to get information that helps them to keep their power. They proceed actually in behind-the-scene negotiations. Therefore, the participation is often only to defend one's own interests. Some powerful actors, like industrialists, abstain from further participation once their interests are safeguarded, mainly because their management of water and wastewater relies upon technical supports (i.e. when their demand is satisfied they often don't see an interest in participating any more since they cannot really get more assets).

The main problem remains that there can be lack of co-ordination or even competition between state administrations at the regional and departmental levels. There can be incoherence in rules and public actions when administrations share the same river. In the case of the Sèvre Nantaise, where the river is the boundary between two Departments, you can take all the water you want on one side, while it is forbidden on the other side. (Isabelle Verdage, Jean-Marc Dziedzicki & Corinne Larrue, Sèvere Nantaise case study)

[1] Almost always the introduction or the increase in valuation of the protection of the environment and nature are part or even the core of the extent changes. Sometimes new human uses like tourism are the extra issues that are taken into account. Where ecological values were already incorporated, new issues might arise and be incorporated, like diffuse agricultural pollution.

[2] The changes in the internal coherence of public governance in most cases included aspects of all five element of public governance: levels and scales, actors and networks, perspectives and objectives, strategies and instruments, and responsibilities and resources for implementation.

The *internal coherence of the property rights* is the degree to which the interdependencies in the water system and its management that occur in reality, are reflected within and between the property and use rights. The essence of this variable is that property and use rights of the one do not inherently or under the given circumstances cause rival uses to affect unavoidably the sustainability of the resource, without external intervention.[3] With the internal coherence of the property rights the picture is somewhat more differentiated. In two cases no improvement or even new inconsistencies occurred. But there were also four cases with a rather complete (change to) coherence in this respect. Generally when absolute limits of the resource are at stake (water, fish) the property and use rights are more used for self-regulatory regimes, than when the protection of the quality of the resource (water, landscape, shores) is at stake. For the water resource in a stricter sense this means that predominant protection by property and use rights occurs more in the 'dry' cases than in the 'wet' cases. In 'wet' cases property and use rights are often restricted and must give way to public governance in order to improve the sustainability of the resource use. At least, this is observed to be common practice.

The *external coherence between public governance and property rights* is the degree to which the interdependencies in the water system and its management that occur in reality, are reflected in the interdependencies between public governance and the property and use rights.[4] The external

[3] Here, for instance, developments were reported like the transfer of shares in relevant private and public companies, privatisation, gradual acceptance of the water body as a common good, lack of introduction of concession system with new uses, introduction of tradable fishing rights, multi-level issues like state ownership as a basis to allow new uses (e.g. to issue gas drilling concessions), while provinces and municipalities hold the public authority to protect other uses, the redistribution of property and use rights, like disposition rights, the buying of land by a user or a public authority to solve conflicting property and use rights, expropriation for similar reasons (rarely and sometimes on the basis of 'expropriation agreements', as in Spain), regulatory unification of the property of land and water, the organisation of users, the acknowledgement of traditional and 'de facto' use rights of some users, agreements (between fishers and kayakists or irrigators and fishermen) to share water use and the withdrawal of informal use rights.

[4] Here the following developments were reported, among others: expired use rights were gradually transferred to other (public or semi-public) institutions, the aim of a minimal water flow is incorporated as a sort of use right for environmental protection, an EU inspired programme gives compensation to farmers for not exerting their use right to part of their farm land, some technical measures require new responsibilities and resources for implementation that demand changes in property rights, adaptation of use rights to public policy aims, voluntary restrictions of the property right holder accepting public policy aims (one of the Belgian cases), the localisation of drinking water industry is problematic but not really considered as a question per se, subsidies allow the regional administration to influence nature management by owners, modification of property rights by creation of

coherence between public governance and property rights changed considerably in half or the cases for the better and in the other half only modestly or less.

The *overall assessment of the regime change* is clearly in most cases that there were considerable improvements on many of the important aspects. Nevertheless also seven worse and 3 better situations occurred.

While 7 of the 12 areas studied were analysed as single cases and the other 5 split into 17 subcases, one might suspect that the subcases are on average more coherent than the single cases, while each subcase only deals with a part of what is relevant. So we tested whether such an artificial 'coherence' bonus was indeed observable in the assessments. This was hardly the case. The assessments of the internal coherence of public governance, the internal coherence of property rights and the external coherence were almost the same with the single (un-split) cases as with the subcases in the sample of 24.[5]

An example of broad improvements

In the Matarranya river process, there are clear signals of regime change, both regarding extension and coherence of the water regime. The extension of the water uses increases as it includes irrigation, population supply, cattle rising, nature protection and tourism. Rivalries between users can be interpreted in territorial terms (intra-basin driven rivalries). There is also an increase of public governance coherence, as it regards levels and scales, multilevel interaction and networks. The most relevant event proving the increase of governance coherence is the Water Agreement reached by the main actors operating at the river basin level. This agreement is the outcome of a process in which a wide range of actors operating at different scales of governance interact: the regional government promoters environmental initiatives; local actors appeal to EU regulation as a legal resource by local actors; the Central Union of Irrigation Communities is created as a body representing all irrigation communities at the basin; PLADEMA -- an ad hoc local association --

zones that are liable to flooding, concessions given by law to user communities, a policy plan to improve the information for self-governing user communities by the development of a census to prevent free riders and by studies, creation of (semi-)public bodies or platforms where practically every user is represented, policies opening up to take also other users than those with a use right to the water itself into account (tourists, fishermen, nature), incorporation of relevant use right holders (farmers, tourists) as targets in public water policy.

5 In two sub-cases the main issue was the purification of wastewater. Since there is a very clear and proven technical solution (building and exploiting treatment plants) all kinds of coherence aspects were hardly problematic when the political will to invest was present. Though this might be seen as an artefact of splitting the case into sub-cases, it can also and probably more rightfully be viewed as a characteristic of situations where a technical solution is possible (though the *siting* of these plants often *does* involve complex interactions). In the other cases there was a lot more social interaction involved and hence relevance for the coherence of the regime.

> aggregates and mobilises actors against the construction of hydraulic works; the Ebro river basin administration negotiates with the local irrigation communities; and the Ministry of Environment finances the construction of lateral pools. These actors, especially those located at the river basin, share a perception of risk caused by an extreme situation of drought among the basin actors and progressively adopt a new water culture.
> Regarding the internal coherence of property rights, some improvements can be identified: the Ebro river basin Plan establishes water needs and uses as well as a minimal ecological flow; some maladjustments between legal aspects and real practices of the CHE and the Central Users Community increase its level of influence regarding decisions on the watering out of the Pena dam and the distribution of water; traditional use rights of some users are respected; and a kind of de facto use rights are given to illegal users of water by the Irrigation Communities of the basin. After the signature of the Water Agreement, the external coherence between public governance and property rights improves to a certain extent. All the main water users have proved to be able to negotiate and reach an agreement based on a common perception of the river as a key element for the future development of the basin. (Meritxell Costejà, Nuria Font & Joan Subirats, – Matarrana River case study)

The general assessment of regime change is as expected positively correlated with all the four aspects, though public governance and external coherence seem to have made the strongest difference for the total assessment.[6] Also most aspects are correlated among each other, though less strong and the least so with the extent variable. This makes sense since it is not one of the set of coherence aspects.

The conclusion on regime change is that in this study in most cases considerable improvements were signalled. But there were also several occasions of more or less failed attempts to regime change.

9.3 Implications of regime changes for sustainable use

The overall sustainability of the resource use was beyond our capacity as social scientists to judge. Furthermore we were especially interested in the

6 Correlations (Kendall's Tau b) with the degree of regime change: (1) extent .481, $p = .004$, (2) coherence of public governance .674, $p = .000$, (3) coherence of property rights .498, $p = 002$, (4) external coherence .741, $p = 000$. If one treats the variables for a moment at interval level and uses multiple regression to estimate the relative contribution to the overall assessment of the four aspects, the regression formula "explains" three quarters of the general assessment of regime change and is statistically significant (adjusted R^2 of .767 with significance $p = .000$). The Beta coefficients of public governance and external coherence are significant. They are .417 and .457 respectively. The extent (.100) and internal coherence of property rights (.072) are statistically not significant and their removal even slightly increases the adjusted R^2 (to .782) with Beta's for the remaining factors of .492 and .511 respectively.

effects of the observed regime changes. This starting point is also part of the solution to the first problem. So the assessment was concentrated on the *implications of the observed regime changes* for indicators that are relevant to sustainability. That is also the reason that the variable is so called. Developments in sustainability of use that clearly have nothing to do with the observed regime changes, but for instance with climate change or rapid economic development are excluded from the judgement.

Further, the weighting between the environment, natural resource protection and risk avoidance on the one hand and the economic and social implications of these ecological changes and/or the measures taken to achieve them on the other hand is a hard nut to crack. We weren't inclined to judge in favour of an increased sustainability without some ecological improvements, even though economic or social indicators might have improved. Here we also paid attention to the relevant EU 'good status' indicators.

> **Rivalries and ecology**
> In the Idro Lake and Chiese River case the problem generates from conflicting interest of the various users of the lake and the water basin. The conflicts occur between water uses for agriculture, hydropower production, tourism, ecological balance, and protection from risks related to flooding, soil erosion, and land sliding. As a response the use of water was managed not only accounting for water needs, but also for water availability. Environmental and land conservation was supported by the maintenance of a constant minimal vital flow, even in summer and controlling the speed of lake depletion. The maximum water-storage level was reduced to avoid the risk of flooding. (Bruno Dente & Alessandra Goria, Idro lake and Chiese River case study)

Often the picture for the *economic consequences* is somewhat mixed. As negative economic consequences the financial costs and/or restrictions for the sectors involved (agriculture, fishery, resource extraction or industry) and in some cases higher water prices are mentioned. On the positive side the following economic phenomena were also often mentioned: gains for tourism, avoidance of future costs, job creation and job safeguarding, and an improved natural resource basis for further economic development. Sometimes lower water costs and increases in productivity were also reported.

> **Nature reserves**
> In the Dender basin, the structure of the economy is modifying. The relative importance of industry and agriculture diminishes as tourism is increasing. In this context changes in the ownership of land are occurring. In fact, associations for the protection of nature buy land to the farmers. Their purpose is to develop natural areas, creating 'green corridors' throughout the region. This activity was initiated and is still supported by the Region. The Flemish Region subsidies the acquisitions. Nature associations negotiate with individual farmers. The farmers are often aged and then get additional financial

resources (to the pension). The two groups of actors benefit from the subsidies of the Region that still manages the conduct of the policy. (David Aubin & Frédéric Varone, Dender River Basin case study)

Tourism development in the Vesdre basin
The low quality of the Vesdre creates rivalries. Pollution prejudices the development of tourism, the only economic reconversion expected for this former industrialised area. At the same time purification of urban wastewater has come compulsory. The tourist sector and the water purification sector are mutually supportive. In both cases the European Union plays the role of institutional interface. In the first place it allocates structural funds. The valley of the Vesdre is classified as an area in economic reconversion. Both tourism development project and purification plants benefit from the subsidies. In the second place, the EU compels the Member States to purify domestic wastewater. As a consequence, the competence authority, i.e. the Walloon Region, developed an ambitious catch up policy and raised the necessary funds. The Vesdre river basin is one of the main recipients. This context should allow of tourist activities in the valley to take off. (David Aubin & Frédéric Varone, Vesdre River Basin case study)

While the economic consequences were mixed, the *social consequences* were often very positive and remarkably varied. The only negative social consequences mentioned were a limitation of land ownership rights and a negative impact on the landscape, both mentioned once. By contrast, the positive social consequences include: modernisation of agriculture, development of new associations of people, more open public debates and more information for the people in general, improved feeling of safety, stop to decline of population and maintenance of young population, fairer distribution between upper and lower communities, resolution of conflict in the local area, improved living conditions, and the reinforcement of the qualities of the river as a key element of social identity.

Concertation
In Wallonia, the tributary basin of the Hoëgne-Wayai hosted a conflict between the fishers and the local mineral water producer. Fishers were complaining about accidental discharges of caustic soda that caused fish disease. During the case, the actors exchanged violent arguments via the press. In order to come out of the conflict, the fishers' federation proposed to the mineral water producer to make a river contract. The river contract is a non-binding, voluntary local concertation mechanism. All the local actors meet and discuss their problems. A monitoring network is put in place. The rivalry is broadened to the whole range of uses. All the quality aspects are taken into account. However, every action is done on a voluntary basis by the actor concerned actor and at its own expenses. Even if results in terms of water quality are mitigated, the initial conflict moved into cooperation and then every local water actor adopted a resource logic. (David Aubin & Frédéric Varone, Vesdre River Basin case study)

Our expectations (expectations 5 and 6) regarding the relation between the regime (change) and the sustainability of institutional resource regimes were:

5. Regimes with a deficient *extent* will be more likely to lead to degradation of water resources or inability to protect the ecological functions of the water resource, than regimes with a larger extent.

6. Regimes with a large extent, but with low *coherence* will be more likely to lead to degradation of water resources or inability to protect the ecological functions of the water resource, than regimes with a similar extent but a higher degree of coherence.

Indeed, the relation between the extent and the sustainability estimates is rather weak and hardly significant, if one leaves out the coherence of the regime aspects.[7] The relation between the general assessment of regime change and the assessment of sustainability is however much stronger.[8] In a scatterplot this is made visual (see Figure 9.1). Remember that several (sub) cases share their values in this plot. This is made visible by the size of the dots.

Figure 9.1: *Relation between the general assessment of regime change and the assessment of sustainability*

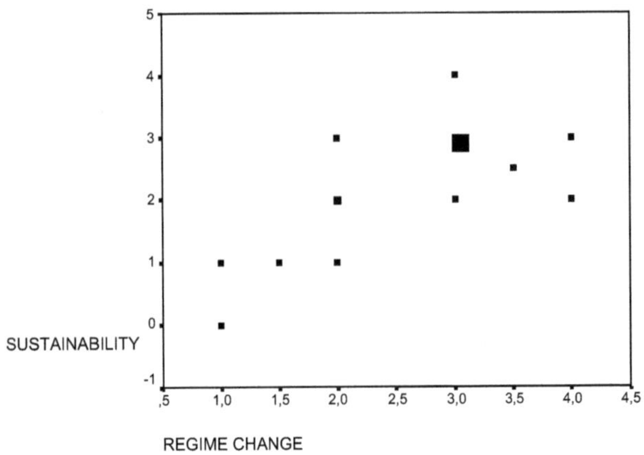

[7] Kendall's Tau b is .297 with one-tailed sign. p = 0,051 and Spearman .342 with sign. p = 0.051. A possibility could be that the relation between extent and sustainability becomes apparent when aspects of coherence are kept constant. When we control for regime change in general the partial correlation is however still weak (.249 with p = .126). Controlling for the three separate coherence aspects together gives about the same result (.176 with p = .223).

[8] Kendall's Tau b is .459 with p = .005 and Spearman .533 with p = .004.

Of the separate regime aspects, by far the most important factor was the coherence of public governance. It correlated even more strongly with the assessment of sustainable resource use than the general regime change.[9]

> **Sustainability and regime changes**
> Regime changes in the case of the Mula river have some positive impacts on sustainability including the environmental, economic and social dimensions. Regarding the environmental dimension, energy and water savings are considerable, there is a decrease in water loses, some measures to avoid the overexploitation of wells and aquifers are adopted, and a minimal ecological flow is established. Regarding the economic dimension, the price of water to farmers is lower than it used to be and the productivity of the huerta improves. Finally, regarding the social dimension, there are some training programs for farmers and an improvement of life quality. In general terms, the positive impacts on sustainability seem to be more related to the increase of internal and external coherence rather than to the increase of extent. (Meritxell Costejà, Nuria Font, Anna Rigol & Joan Subirats – Mula River case study)

To summarise the results:

A scatterplot illustrates the relation between the coherence of public governance and the assessment of the sustainability of the water resources regime (see Figure 9.2). The conclusion is here that there is only weak support for our first expectation: that an increased extent contributes as such to a more sustainable resource use. The support for the second expectation -- that increased coherence contributes to a more sustainable resource use -- is much stronger. Though this can be regarded as supportive evidence for the proponent of 'integral water management', it should be considered that this isn't a sort of 'mechanic' causal relationship. It still holds true that 'the devil is in the details' (expectation 4).

9 Kendall's Tau b of the correlations with the sustainability of resource use are with coherence public governance: .635, p = .000, with the coherence of property rights .457, p = .005, with the degree of external coherence .326, p = .035, and with extent 297, p = .051.

Figure 9.2: *Relation between the coherence of public governance and the assessment of the sustainability of the water resources regime.*

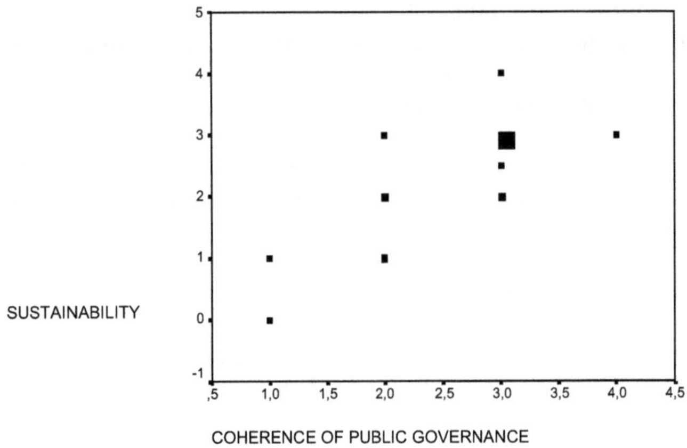

Voluntary restriction
The lower part of the Vesdre river basin was regularly under water due to water releases from the dams of Eupen and the Gileppe. During periods of heavy rainfalls, the dam reservoirs reached their maximum capacity and it became dangerous to stock more water. People and communes downstream were complaining. Consultations went on to circumscribe the problem in the basin despite the lack of regulation. In fact the manager of the dam agreed with the main user of the reservoir, i.e. the drinking water producer, to constitute a higher safety margin in case of significant rainfall. The two actors have endorsed the risk of water shortages in drought periods. The dams no longer threaten the downstream part of the basin. Informal agreements were later extended to other consequences of water releases, i.e. minimum flows and extraordinary releases for canoeing. Moreover, this kind of agreement generated an extended mobilisation of all the local actors involved in water quantity management as the problem of floods remains, but on different patterns. (David Aubin & Frédéric Varone, Vesdre River Basin case study)

9.4 Explaining regime changes by change agents and conditions

Change agents
This is the combined force of the listed change agents as an impetus to set in motion regime changes in the direction of more integration. The joint force of the identified change agents in the cases is often assessed as considerable. Only in 7 (sub)cases is it assessed as rather weak. The types of change agents mentioned were EU originated pressures, national regime

developments, problem pressures and various other case circumstances. In 13 of the 24 cases EU policies were mentioned as relevant.[10] In all but two cases national policy and regime changes were.[11] In only five cases was there not much problem pressure.[12] In 10 cases various other circumstances were mentioned.[13]

> **Example of a set of change agents on case level**
> Change agents in the case of the Mula River include the leadership of regional government, which has technical and financial resources and support from other institutions (EU, national administration) in the elaboration of the Modernisation Plan. Of crucial importance is the ability of the Irrigation Community to break the Heredamiento monopoly of water distribution. Problem pressure also becomes an important change agent -- drought conditions precipitate a deep crisis of the traditional structure of the Mula huerta. In addition, policy initiative and new scientific knowledge about the state of the resource are important variables leading to a regime change. (Meritxell Costejà, Nuria Font, Anna Rigol & Joan Subirats – Mula River case study)

Maybe national policy support is a necessary, though not a sufficient condition.[14] The two cases where such influence was not reported had a very

[10] As such a great variety of EU policies were mentioned as relevant: the standard for minimal flow of rivers, (national laws that were triggered by) directives on the water basis system, the 1991 waste water treatment directive (5x), phosphate and nitrate standards, fishery policies, the 1972 wild birds and 1992 habitat directives with their special protected areas (3x), the 1975 drinking water directive (3x) (and the role of the European Court of Justice to force implementation), the regional development policy with its structural funds (2x). More generally various EU regulations were used as arguments in the debates, even when not self-enforcing.

[11] Apart from various 'normal' water (and some nature) policies, some more regime oriented pressures were also mentioned: promoting regime development at the level of the water basins (3x), laws demanding (land-use) planning (4x), acts that allow the government as owner of the water to regulate fishing on the basis of considerations of nature protection, environmental impact assessment, white papers pushing for 'integral water management' (3x), federalisation (Belgium), legislation allowing expropriations and indemnities in favour of flood protection, and the designation of parts of the basin as nature protection area. Note that several of these are not or might not be independent from the relevant EU policies!

[12] With the problems at hand there is a clear division between 'wet' cases (the majority) and 'dry' cases. In the dry cases increased use by agriculture and tourism are main problem causes. In the wet cases pollution and the risk of flooding are the most mentioned problems. For almost all cases the increased value attached to nature and environment considerations makes these enter the picture as 'new' problem pressures.

[13] Some examples are: the expiration of concessions for irrigation, changing market regulations pushing for new economic developments, state withdrawal from participation in economic developments, expanding land use for building, the break-down of traditional management regimes, experts providing new information, local and environmental associations and devoted individuals.

[14] Often the national government provided crucial resources like formal rules and money.

low overall force of change agents. But generally it is not the type of change agents or the presence of a variety of them that matters. Each change agent can 'do the job' of exerting a major 'force of change agents' if it is pressing enough.

Our expectations (expectations 1 and 2) regarding the relation between the general force of the change agents and regime change were:

1. Most change agents (in the period and context of our cases) will lead to more differentiation in the regime (resulting in more complex regimes with a higher degree of extent).

2. Other external change agents of a specific nature can also lead to coherence in or between one or some elements of the regime, but only in combination with deliberate attempts of motivated actors (ultimately resulting in coherent regimes or in 'failed' regime shifts with encapsulated initial changes).

As expected (expectation 1) of the various forms of regime change, only the extent seems directly related to the force of the change agents[15]. For the other relations more is necessary. And these attempts to attain more coherence are expected to depend on several conditions.

Finding political will

In Verviers, drinking water consumption has led to lead-poisoning for more than a century. Poisoning was due to lead pipes attacked by naturally acid water. Diverging interests and the weakness of knowledge around the nature of the contamination explained the status quo. The dam that provides water to the town had initially been build for industrial uses. The network was later extended to private housings and water declared to be drinking water without prior treatment. Acid water was very convenient for the industries because of its cleaning properties. This position was well reflected in the municipal council. The commune was the owner of the water distribution service. In 1980, the EU drinking water directive set up constraining standards for lead concentration in drinking water. The commune of Verviers had to adapt but missed both the political will and the financial means. Finally, the building of a treatment plant was taken in charge by the Region and a deviation of the main pipe did not counteract the industrial interest. Work began only when industry had guarantee on the unchanged properties of its water. The public health problem was taken into account without inducing any redistribution at the detriment of other water uses, industry in the present case. (David Aubin & Frédérich Varone, Vesdre River Basin case study)

Bottom-up regime changes

Sometimes it was not national regime change influencing the extent of the regime at the case level, but the other way around. Here are two examples of

[15] Kendall's Tau b: correlation of force of change agents with: extent .393, p= .015, coherence public governance .112, p = .270, coherence property rights .067, p = .349, external coherence .127, p = .237, and general regime change .171, p = .168.

> bottom-up processes and subsequent 'legitimisation' of local developments through national legislation in Switzerland.
> The process of regional regime inventions arising from local problem pressure which are subsequently supported and thus legitimated by changes in the policy design at federal level can be observed in both Swiss case studies. In the Seetal valley, the canton of Lucerne had already issued a notice in 1988 reducing the restrictions on the number of production animals on farms from four to three livestock units per hectare. Even if this restriction was never really implemented at regional level, it served as a model for the introduction of the same restriction into the Federal Law on Water Protection of 1991. In the Maggia valley in the canton of Ticino, quantitative protection of the water resources dates back to 1976, anticipating the changes in the federal regime by a wide margin. At the level of the water basin, protective measures in terms of minimal residual flows were applied in 1982, a full 10 years before the enactment in the Federal Law on Water protection of 1991. (Corine Mauch & Adèle Thorens – Swiss case studies)

Conditions

This is the degree to which the listed conditions provide, separately and as a set, favourable or unfavourable conditions for regime changes in the direction of more integration (extent and coherence).

Expectation 3 was that attempts to change regimes into a more integrated status would have relatively more success when:

- There is already a longer tradition of thinking in terms of co-operation in the water management sector or such a thinking is built during the case early enough to influence later stages of the case history.
- There is a common understanding that the counteracting (side) effects of non-integrated water management harm sustainability and that this sooner or later will have to be stopped anyhow (joint problem).
- There is a notion of possible joint gains from coherence, so-called 'win-win situations' (joint opportunities).
- There is a credible threat of a (potentially) dominant actor accumulating power and altering the public governance pattern in his own way and to his own interest when no solution is reached (credible alternative threat).
- There are well functioning institutions that provide fertile ground for coherence attempts (institutional interfaces).

Generally the researchers assessed that in their (sub)cases there was no very stimulating tradition of earlier co-operation between the actors involved in the rivalry/ies. Joint problem awareness has been present to some extent in several cases, though often only on a part of the relevant aspects or only with some of the relevant actors. There has been considerable differentiation between the cases in terms of the degree to which the actors involved saw chances to actually gain together from solving the rivalry by a more integrated regime. In one case there was even a sense of joint loss. With the condition of a credible threat of interventions by a dominant actor to solve the disputes to his own benefit too there has been a considerable differentiation among the cases. Generally speaking the condition of institutional interfaces was somewhat better that most of the other conditions. Nevertheless, in many cases these were only on a part of the relevant aspects or not functioning very well.

All in all, in many cases the assessments of the conditions for regime change taken together are regarded as rather favourable. In nine cases the conditions are viewed less favourably. Especially the awareness of joint chances and good institutional interfaces -- and to a lesser extent an existing tradition of co-operation was seen as important positive conditions for regime change.[16] Lower assessments of the general conditions indeed correlate with smaller regime changes, as expected in expectation 3.[17] Figure 9.3 shows this relationship in a visual way.

Figure 9.3: *The relation between conditions and regime changes.*

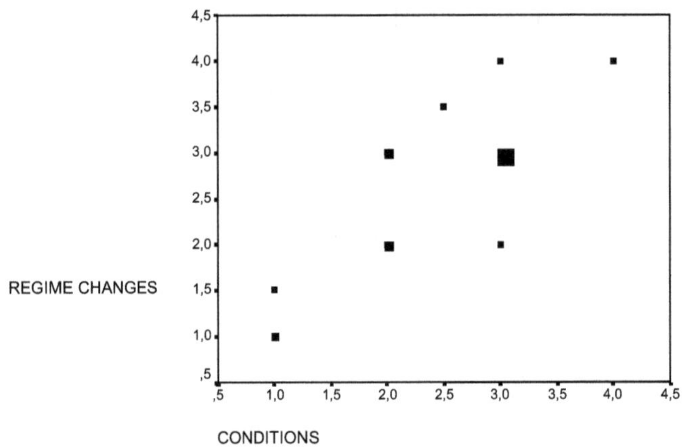

Polders and wateringues versus water floods

All along the river Dender in Flanders, riparian landowners are involved in a particular kind of public administration, the polders and the wateringues. The polders and wateringues manage drainage on their territory. They finance their activity with direct taxation. Draining activities are in conflict with the need to create buffer zones. The competent authority for water quantity management

16 The correlations (Kendall'stau b) of the general assessment of the conditions and the separate conditions is as follows: Tradition of co-operation: .365, p=.022, Joint problem awareness: .194, p=.138, Joint chances: .457, p=.005, Credible alternative threat: .148, p=.202, Institutional interfaces: .611, p=.000.

17 The correlation is Kendall's tau b .633 with p = .000 and Spearman's rho .687 with p = .000.

> on the Dender faces frequent floods of growing importance. As the problem pressure is growing, solutions introduced are residual. The weakness is due to an absence of common concern between the involved users. The water manager has no possibility to build new relief basins. It only builds dikes to divert the flood. It is not confronted with claims from the polders and wateringues that come under flow as a matter of tradition. Moreover, there is no real mechanism of concertation between the users and no coordination between the various competent authorities. Mutual information remains weak. Everyone is only preoccupied by the evacuation of water out of its territory. The problem should increase without a sustainable response is put in place. The only answer consists in building a huge pumping station at the mouth of the Dender. (David Aubin & Frédéric Varone, Dender River Basin case study)

The research expectation suggests possible different effects of different conditions being more or less favourable. We investigated this by correlation analysis between all the conditions and all the aspects of regime change.

The *general assessment of the conditions* not only correlated with the general regime change, but also with all other aspects of regime change.[18]

The *tradition of co-operation* in the water management sector showed only significant correlations with the internal coherence of public governance and almost with the extent.[19]

The condition of a common understanding that the counteracting (side) effects of non-integrated water management harm sustainability and that this sooner or later will have to be stopped anyhow (*joint problem*) did not correlate significantly with any of the regime indicators. Closest was the correlation with the internal coherence of public governance (tau .273, p = .067).

The notion of possible joint gains from coherence, so-called 'win-win situations' (*joint opportunities*) again correlated with some regime indicators.[20] The fact that the correlation of the awareness of 'win-win' situations is even stronger with the external coherence between public governance and property rights than with the general regime change is striking. It might point to an often 'public-private' nature of such opportunities.

The credible threat of a dominant actor accumulating power and altering the public governance pattern in his interest when no solution is reached (*credible alternative threat*) did not correlate significantly with any of the

[18] With the extent (tau .658, p = .000), the internal coherence of public governance (tau .749, p = .000), the internal coherence of property rights (tau .514, p = .002) and the external coherence between public governance and property rights (tau .469, p = .005).

[19] With internal coherence of public governance tau .360, p = .026, and with the extent tau .277, p = .064.

[20] With the internal coherence of public governance (tau .452, p = .007), the external coherence between public governance and property rights (tau .531, p = .001) and with the general regime change (tau .486, p = .003).

regime indicators. It almost significantly correlated with the extent, though (tau .261, p = .071).

Last but not least, the existence of well functioning institutions that provide fertile ground for coherence attempts (*institutional interfaces*) correlates with the extent (tau .668, p = .000), internal coherence of public governance (tau .397, p = .014) and with the general regime change (tau .339, p = .027).

So all in all, of the separate conditions (and the force of change agents) the *joint opportunities* and the *institutional interfaces* conditions stand out in the explanation of the various forms of regime changes.

Joint opportunities and institutional interfaces around the IJsselmeer
Sometimes rival uses can nevertheless be turned into win-win situations. The shores of the Dutch IJsselmeer (Lake IJssel) have rival uses of inter alia nature (bird habitat) and tourism (boating marinas). Of course tourism on the other hand benefits from beautiful nature. Seeking the balance between the two uses can therefore be beneficial for both. With a homogeneous use like IJsselmeer fisheries there is rivalry between the users, but on the other hand all users have a certain interest in a just distribution of rights, and therefore may favour a regime that guarantees this while preventing a 'tragedy of the commons'. This creates a basis for joint action that can be further exploited by having the right institutional interfaces in place.
These institutional interfaces can be triggered by European and national measures. In the IJsselmeer case the national government founded a negotiation platform, a steering committee on the so-called corner lakes, a producers' organisation on fishery, environmental impact procedures (gas drilling) and land use planning procedures with open participation. Such institutions catalysed the involvement of users and other citizens (cf. the EU WFD) and functioned sometimes as 'policy brokers' and sometimes as forms of 'institutional leadership'. (Dave Huitema, IJsselmeer case study)

9.5 Outlook: our conclusions in the perspective of the European water management policy

European water policy has developed along two lines -- water quality and emission standards -- that reflect different national views. The new European Water Framework Directive (WFD) is an attempt to reconcile the two approaches and to integrate water quantity aspects. The purpose of the WFD is to achieve a good ecological quality for all waters inside the European Union, at the scale of water basins, where an authority implements integrated management programmes. The WFD should guarantee, as of 2015, a 'good status' for all ground- and surface waters, in quality and quantity, according to an eco-centred logic. In order to achieve this goal it promotes an integrated water management, i.e. a management that considers all the water aspects and legislation in a single picture and on a delineated territory, the water basin. The integration of control and action should occur for quality and quantity aspects, surface and groundwater, exploitation and preservation, objectives of quality and emission limit values and water policy vis-à-vis other policies. The WFD sets up guidelines and leaves significant room for manoeuvre to the Member States. The guidelines allow an evaluation and a comparison of the efforts developed by the Member States and their results.

The main concepts of this book are in close relation to the central themes of the new European water policy. The *'good status'* of the WFD is related to the ultimate dependent variable in our analysis, the degree of 'sustainable use', especially to the ecological aspect of sustainable use. However, even in the 1996 communication leading to the WFD due attention is also paid to the 'evaluation of costs'. This aspect is reflected in the 'economic consequences' aspect of sustainable use. We observed that besides costs other negative, but often also positive economic consequences could be observed. A third aspect that was included in our research was that of the social consequences. Here a remarkable number of positive developments were reported from the case studies. Generally, a higher degree of sustainable use correlated with a more integrated regime at the water basin level, just as was expected by both the theory described in Chapter 2 and the 'practical policy theory' underlying the WFD. Though this can be regarded as supporting evidence, it should be considered that this isn't a sort of 'mechanical' causal relationship. Under certain circumstances it can even be envisioned that more integration leads to deterioration of sustainable use. It still holds true that 'the devil is in the details'. Nevertheless, empirically in our 24 cases the relationship between integrated management and the status of the water resources shown to correspond with the ideas guiding the WFD.

The main venue by which the new European water policy seeks to improve the good status of European waters is by 'integrated water management at water basin scale'. In this book the cases that are studied were not at the full water basin scale, but at the lower level of tributary river basins. The reason for this is that we believe that integration of management is a multi-level endeavour. At the higher level of international rivers like the Rhine or even large national rivers like the Loire, circumstances vary to such a degree that there is not one, but several sets of uses and users and consequently also multiple resource regimes needed at a sub-basin level. This is not to state that the full water basin should not be in need of coordinated management, but only that for impacting many uses and users, sub-regimes at a tributary river basin level are also needed. This idea is in accordance with the principle of subsidiarity that is explicitly endorsed in European water policy. The case studies concentrated on this level (with areas of some 500 to 2500

km^2) and found many interesting experiences with (attempts to achieve) more integrated water management. These illustrate the assumption of the European water policy that it is necessary to accept some variation of the institutional arrangements that are used to promote integrated management. Though the organisation of management on a sub-basin level is left predominantly at the discretion of the member states, we think that at least devices for Europe-wide communication and exchange on experiences with integral water management on that level could be helpful for the actual practical implementation of the WFD. This could be part of 'joint implementation' arrangements.

Integrated water management in this book is conceptualised with the help of the concepts of extent and coherence. The 'extent' of the regime reflects the elements of integration in the WFD that stress that all relevant directives and all waters in the area should be managed in a combined approach. In this book we stress the completeness of the regime to regulate all relevant uses and users. The elements that stress multi-level (even international if necessary) and multi-actor (stakeholders and citizens) involvement and the coherent action guided by management plans are reflected in the concept of 'public coherence'. As a special feature of our research, not only the coherence of public governance, but also the coherence of the property & use rights regime and the coherence of the relation between public governance and property and use rights are included in the assessments. The study illustrates that these are important aspects of the water management regimes, especially -- but not exclusively -- when quantity issues are at stake. Theoretically it can be expected that inclusion of former socialist economies in Eastern Europe would increase the variation in the regimes of property and use rights considerably and would make this issue even more important. In Switzerland public policies that reduce use rights by more than 7% need to include compensations acknowledging these rights. All aspects of integrated water management studied seem to make a difference, though not equally in all cases. The research in this book has shown that special attention to the property and use rights affected and the relation between those and the public governance measures is a worthwhile extension of the focus of integrated water management.

The integration between water management and other sector policies is in the new European water policy envisioned by the mechanisms embedded in 'full cost pricing'. In our cases we did not specifically encounter this subject. Consequently we don't have a conclusion on full cost pricing. But what we did encounter were a number of cases in which other than direct water issues entered the process of development of new water regimes. Examples are issues of landscape, wetlands and fishery, which were entered into the debate by interested actors. Though 'full cost pricing' could be important to send the right price signals to all actors, there will probably remain various rivalries that need a form of integrated water management that deliberately tries to bridge externally to other sector policies for coordination.

The research in this book did spend a great deal of effort in getting a better insight into a variety of change agents and conditions that stimulate more integrated water management. For we learned that integrated management regimes are not something that one can 'proclaim into reality'. Deliberate attempts by motivated actors are surely needed to realise it in practice. We won't repeat all our conclusions on this subject here, but concentrate on the points where EU policies come in.

Among the change agents we have seen that in more than half of the cases EU directives and other policies play an important role. Among these directives are also some that are not directly 'water directives'. Another observation is that national policies that are mentioned as leading to regime changes were often in their turn triggered or in any case related to EU directives.

Even more important than the change agents mentioned proved to be the conditions for change. The European Union can have important -- indirect -- effects here too. A first observation is that European policies are often used in the internal debate at case level as arguments to pursue a certain position. This holds especially for NGOs and other actors with little formal power and of course when they want to move in the same direction as the relevant EU policy involved. Even when these policies are non-obligatory, in this way they have a certain influence. Of course, part of this influence is generated by the prospect that these policy lines will become more compelling after a while. So for the WFD aim of participation in water management, EU policies can play an important role. Of the several conditions joint chances and institutional interfaces proved to be the most important. Both can be seen as venues at which to aim supplementary EU measures in the context of joint implementation, to improve the chances for regime changes in the direction of integrated water management.

by interested actors. Though 'full cost pricing' would be important to send the proper price signals to all actors, there will probably remain various rivalries that need a form of integrated water management that [illegible] types to tackle externally to their actions [illegible] coordination.

The research in this book did need a great deal of effort in gathering better [illegible]

ENVIRONMENT & POLICY

1. Dutch Committee for Long-Term Environmental Policy: *The Environment: Towards a Sustainable Future.* 1994 ISBN 0-7923-2655-5; Pb 0-7923-2656-3
2. O. Kuik, P. Peters and N. Schrijver (eds.): *Joint Implementation to Curb Climate Change.* Legal and Economic Aspects. 1994 ISBN 0-7923-2825-6
3. C.J. Jepma (ed.): *The Feasibility of Joint Implementation.* 1995 ISBN 0-7923-3426-4
4. F.J. Dietz, H.R.J. Vollebergh and J.L. de Vries (eds.): *Environment, Incentives and the Common Market.* 1995 ISBN 0-7923-3602-X
5. J.F.Th. Schoute, P.A. Finke, F.R. Veeneklaas and H.P. Wolfert (eds.): *Scenario Studies for the Rural Environment.* 1995 ISBN 0-7923-3748-4
6. R.E. Munn, J.W.M. la Rivière and N. van Lookeren Campagne: *Policy Making in an Era of Global Environmental Change.* 1996 ISBN 0-7923-3872-3
7. F. Oosterhuis, F. Rubik and G. Scholl: *Product Policy in Europe: New Environmental Perspectives.* 1996 ISBN 0-7923-4078-7
8. J. Gupta: *The Climate Change Convention and Developing Countries: From Conflict to Consensus?* 1997 ISBN 0-7923-4577-0
9. M. Rolén, H. Sjöberg and U. Svedin (eds.): *International Governance on Environmental Issues.* 1997 ISBN 0-7923-4701-3
10. M.A. Ridley: *Lowering the Cost of Emission Reduction: Joint Implementation in the Framework Convention on Climate Change.* 1998 ISBN 0-7923-4914-8
11. G.J.I. Schrama (ed.): *Drinking Water Supply and Agricultural Pollution.* Preventive Action by the Water Supply Sector in the European Union and the United States. 1998 ISBN 0-7923-5104-5
12. P. Glasbergen: *Co-operative Environmental Governance: Public-Private Agreements as a Policy Strategy.* 1998 ISBN 0-7923-5148-7; Pb 0-7923-5149-5
13. P. Vellinga, F. Berkhout and J. Gupta (eds.): *Managing a Material World.* Perspectives in Industrial Ecology. 1998 ISBN 0-7923-5153-3; Pb 0-7923-5206-8
14. F.H.J.M. Coenen, D. Huitema and L.J. O'Toole, Jr. (eds.): *Participation and the Quality of Environmental Decision Making.* 1998 ISBN 0-7923-5264-5
15. D.M. Pugh and J.V. Tarazona (eds.): *Regulation for Chemical Safety in Europe: Analysis, Comment and Criticism.* 1998 ISBN 0-7923-5269-6
16. W. Østreng (ed.): *National Security and International Environmental Cooperation in the Arctic – the Case of the Northern Sea Route.* 1999 ISBN 0-7923-5528-8
17. S.V. Meijerink: *Conflict and Cooperation on the Scheldt River Basin.* A Case Study of Decision Making on International Scheldt Issues between 1967 and 1997. 1999 ISBN 0-7923-5650-0
18. M.A. Mohamed Salih: *Environmental Politics and Liberation in Contemporary Africa.* 1999 ISBN 0-7923-5650-0
19. C.J. Jepma and W. van der Gaast (eds.): *On the Compatibility of Flexible Instruments.* 1999 ISBN 0-7923-5728-0
20. M. Andersson: *Change and Continuity in Poland's Environmental Policy.* 1999 ISBN 0-7923-6051-6

ENVIRONMENT & POLICY

21. W. Kägi: *Economics of Climate Change: The Contribution of Forestry Projects*. 2000 ISBN 0-7923-6103-2
22. E. van der Voet, J.B. Guinée and H.A.U. de Haes (eds.): *Heavy Metals: A Problem Solved?* Methods and Models to Evaluate Policy Strategies for Heavy Metals. 2000 ISBN 0-7923-6192-X
23. G. Hønneland: *Coercive and Discursive Compliance Mechanisms in the Management of Natural Resourses*. A Case Study from the Barents Sea Fisheries. 2000 ISBN 0-7923-6243-8
24. J. van Tatenhove, B. Arts and P. Leroy (eds.): *Political Modernisation and the Environments*. The Renewal of Environmental Policy Arrangements. 2000 ISBN 0-7923-6312-4
25. G.K. Rosendal: *The Convention on Biological Diversity and Developing Countries*. 2000 ISBN 0-7923-6375-2
26. G.H. Vonkeman (ed.): *Sustainable Development of European Cities and Regions*. 2000 ISBN 0-7923-6423-6
27. J. Gupta and M. Grubb (eds.): *Climate Change and European Leadership*. A Sustainable Role for Europe? 2000 ISBN 0-7923-6466-X
28. D. Vidas (ed.): *Implementing the Environmental Protection Regime for the Antarctic*. 2000 ISBN 0-7923-6609-3; Pb 0-7923-6610-7
29. K. Eder and M. Kousis (eds.): *Environmental Politics in Southern Europe: Actors, Institutions and Discourses in a Europeanizing Society*. 2000 ISBN 0-7923-6753-7
30. R. Schwarze: *Law and Economics of International Climate Change Policy*. 2001 ISBN 0-7923-6800-2
31. M.J. Scoullos, G.H. Vonkeman, I. Thornton, and Z. Makuch: *Mercury - Cadmium-Lead: Handbook for Sustainable Heavy Metals Policy and Regulation*. 2001 ISBN 1-4020-0224-6
32. G. Sundqvist: *The Bedrock of Opinion*. Science, Technology and Society in the Siting of High-Level Nuclear Waste. 2002 ISBN 1-4020-0477-X
33. P.P.J. Driessen and P. Glasbergen (eds.): *Greening Society*. The Paradigm Shift in Dutch Environmental Politics. 2002 ISBN 1-4020-0652-7
34. D. Huitema: *Hazardous Decisions*. Hazardous Waste Siting in the UK, The Netherlands and Canada. Institutions and Discourses. 2002 ISBN 1-4020-0969-0
35. D. A. Fuchs: *An Institutional Basis for Environmental Stewardship: The Structure and Quality of Property Rights*. 2003 ISBN 1-4020-1002-8
36. B. Chaytor and K.R. Gray (eds.): *International Environmental Law and Policy in Africa*. 2003 ISBN 1-4020-1287-X
37. F.M. Brouwer, I. Heinz and T. Zabel (eds.): *Governance of Water-Related Conflicts in Agriculture*. New Directions in Agri-Environmental and Water Policies in the EU. 2003 ISBN 1-4020-1553-4

ENVIRONMENT & POLICY

38. G.J.I. Schrama and S. Sedlacek (eds.): *Environmental and Technology Policy in Europe. Technological Innovation and Policy Integration.* 2003 ISBN 1-4020-1583-6
39. A.J. Dietz, R. Ruben and A. Verhagen (eds.): *The Impact of Climate Change on Drylands*. With a focus on West Africa. 2004 ISBN 1-4020-1952-1
40. I. Kissling-Näf and S. Kuks (eds.): *The Evolution of National Water Regimes in Europe*. Transitions in Water Rights and Water Policies. 2004 ISBN 1-4020-2483-5
41. H. Bressers and S. Kuks (eds.): *Integrated Governance and Water Basin Management.* Conditions for Regime Change and Sustainability. 2004 ISBN 1-4020-2481-9

For further information about the series and how to order, please visit our Website
http://www.wkap.nl/series.htm/ENPO

KLUWER ACADEMIC PUBLISHERS – DORDRECHT / BOSTON / LONDON

MIX
Papier aus verantwortungsvollen Quellen
Paper from responsible sources
FSC® C105338

If you have any concerns about our products,
you can contact us on
ProductSafety@springernature.com

In case Publisher is established outside the EU,
the EU authorized representative is:
Springer Nature Customer Service Center GmbH
Europaplatz 3, 69115 Heidelberg, Germany

Printed by Libri Plureos GmbH
in Hamburg, Germany